Synthesis Lectures on Mathematics & Statistics

Series Editor

Steven G. Krantz, Department of Mathematics, Washington University, Saint Louis, MO, USA

This series includes titles in applied mathematics and statistics for cross-disciplinary STEM professionals, educators, researchers, and students. The series focuses on new and traditional techniques to develop mathematical knowledge and skills, an understanding of core mathematical reasoning, and the ability to utilize data in specific applications.

Aleksei Beltukov

Differential Equations and Data Analysis

 Springer

Aleksei Beltukov
University of the Pacific
Stockton, CA, USA

ISSN 1938-1743   ISSN 1938-1751 (electronic)
Synthesis Lectures on Mathematics & Statistics
ISBN 978-3-031-62256-4   ISBN 978-3-031-62257-1 (eBook)
https://doi.org/10.1007/978-3-031-62257-1

© The Editor(s) (if applicable) and The Author(s), under exclusive license to Springer
Nature Switzerland AG 2025

This work is subject to copyright. All rights are solely and exclusively licensed by the Publisher, whether the whole or part of the material is concerned, specifically the rights of translation, reprinting, reuse of illustrations, recitation, broadcasting, reproduction on microfilms or in any other physical way, and transmission or information storage and retrieval, electronic adaptation, computer software, or by similar or dissimilar methodology now known or hereafter developed.
The use of general descriptive names, registered names, trademarks, service marks, etc. in this publication does not imply, even in the absence of a specific statement, that such names are exempt from the relevant protective laws and regulations and therefore free for general use.
The publisher, the authors and the editors are safe to assume that the advice and information in this book are believed to be true and accurate at the date of publication. Neither the publisher nor the authors or the editors give a warranty, expressed or implied, with respect to the material contained herein or for any errors or omissions that may have been made. The publisher remains neutral with regard to jurisdictional claims in published maps and institutional affiliations.

This Springer imprint is published by the registered company Springer Nature Switzerland AG
The registered company address is: Gewerbestrasse 11, 6330 Cham, Switzerland

If disposing of this product, please recycle the paper.

To the giants on whose shoulders we stand.

Preface

Having taught Ordinary Differential Equations (ODE) for nearly 20 years to engineering and science majors, I have come to the conclusion that the first thing they need to learn is modeling real-world data with linear ODE with constant coefficients. That is what this book is about. Nonlinear ODE are, of course, very important and interesting, arguably more so than their linear counterparts. However, nonlinear theory is the *second* thing that should be learned, and only after linear theory has been mastered.

The book begins with simple population modeling examples in Chap. 1 and ends with frequency response analysis in Chap. 10. Hopefully, the eight chapters in between make the elevation gain somewhat gradual, but that depends on the reader's preparation. As a minimum, I expect multivariate calculus, some familiarity with matrices, programming literacy that is sufficient for parsing the included MATLAB code, and a bit of probability theory. Linear algebra will be very beneficial as well.

With the exception of Chaps. 5 and 10, each chapter contains at least one section where physical data is modeled and analyzed. Some chapters contain three or four such sections. In this respect, the book is similar to Martin Braun's *Differential Equations and Their Applications*, published by Springer-Verlag in the early 1980s, but in the following respects it is different.

Firstly, there is a narrow focus on linear theory. This is justified in Chaps. 3 and 4 where it is shown that many ODE that are relevant in science and engineering are either linearized or are linear to begin with. Besides making the book coherent and concise, the exclusion of nonlinear theory created space for topics in linear theory and data analysis that are rarely found in ODE texts, such as the connection between nonlinear least squares and maximum likelihood estimates in Chap. 2, a thorough discussion of vector spaces in Chap. 6, and multidimensional convolution in Chap. 7.

Secondly, data analysis is treated on equal footing with linear ODE theory. It has two dedicated chapters—Chap. 2 on estimation of model parameters using nonlinear least squares and Chap. 8 on discrete Fourier transform (DFT)—and is consistently practiced in other chapters. I tried to include either full or downsampled data sets as tables whenever possible; when I could not find good quality data, I simulated it in MATLAB and included code. Some authors urge their readers to have pen and paper at the ready. I urge mine to have MATLAB (or its equivalent) running on their computer: the mathematics discussed in this book is best learned as it is translated into code.

Thirdly, I abandoned traditional "definition-lemma-theorem" style of mathematical exposition in favor of a more conversational style. My main goal is not to show "how" but to explain "why": Why does one separate variables? Why do complex exponentials make sense? Why are Fourier coefficients what they are? Why does the model deviate from experiment? Also, there are no citations in the main text: all attributions are in the "Comments and bibliography" sections that end each chapter.

Answering the "why" questions is more difficult than answering the "how" questions, often because of misconceptions that stand in the way. Chapter 5 addresses one popular misconception about the exponential function—that it is a strange number e multiplied by itself x times; in my experience, most sophomores have trouble accepting that complex and matrix exponentials make sense because of that. The "linear algebra" Chap. 6 has abstract algebra content aimed at clearing two other misconceptions: that imaginary numbers are "figments of our imagination" and that vectors are "things with magnitude and direction." I took the unusual step of defining number fields to dispel all doubts surrounding complex numbers. That is followed by an abstract definition of a vector space whose purpose is to show that vectors are defined by vector operations rather than their geometric attributes.

While most of the material in the book is standard, the presentation often is not. For instance, the derivation of the convolution formula in Chap. 7 uses only simple Calculus and does not rely on Dirac's delta function—I have not seen this in other books. The exposition of Fourier analysis begins with DFT, presented as a way of approximating functions with trigonometric polynomials, and proceeds to Fourier series, derived as limits of DFT-based approximations, and then to Fourier transform, derived as a limiting case of Fourier series: this may not be a new idea, but it is unorthodox.

In summary, this book presents linear ODE theory and immerses the reader into modeling of real data, with all its warts. By itself, it may not be suitable for a conventional ODE course, however, I hope that it will serve as a useful supplement.

Stockton, California Aleksei Beltukov
April 2024

Acknowledgements The author thanks the staff of Springer Nature for encouragement, technical help, and infinite patience. The author also thanks his colleagues, Drs. Andy Lutz and John Mayberry, and his students, especially Christopher Uchizono, for reading the manuscript and providing invaluable feedback.

Contents

1 Example of Modeling Population Dynamics 1
   1.1   Size-Limited Growth 1
   1.2   Natural Growth Equation 2
   1.3   Separation of Variables 3
   1.4   Use of Initial Values 4
   1.5   Exponential Fit via Linear Regression 5
   1.6   Logistic Equation 6
   1.7   Logistic Fit via the Simpex Method 8
   1.8   Comments and Bibliography 10
   1.9   Exercises 12
   References 12

2 Estimation of Parameters 13
   2.1   Maximum Likelihood Estimation 14
   2.2   Newton's Method in One Dimension 16
   2.3   One-Parameter Exponential Fit 19
   2.4   Newton's Method for Systems 20
   2.5   Two-Parameter Exponential Fit 21
   2.6   Nonlinear Regression in 1-Norm 24
   2.7   Comments and Bibliography 25
   2.8   Exercises 25
   References 26

3 Linearized ODE and Exponential Laws 27
   3.1   Convective Heat Transfer 28
   3.2   The Cooling Light Bulb Experiment 29
   3.3   RC-Circuit 31
   3.4   The Discharging Capacitor Experiment 34
   3.5   One-Dimensional Motion with Resistance 36

	3.6	Usain Bolt's World Record	36
	3.7	Atmospheric Pressure	38
	3.8	NASA 1976 Standard Atmosphere Model	39
	3.9	Electro-Thermal Analogy	40
	3.10	Comments and Bibliography	41
	3.11	Exercises	46
		References	48
4	**Memoryless Processes**		49
	4.1	Radioactive Decay as a Memoryless Process	49
	4.2	Markov Description of Radioactive Decay	52
	4.3	Absorption of Light	53
	4.4	Mutarotation of Glucose	54
	4.5	Comments and Bibliography	57
	4.6	Exercises	58
		References	59
5	**Exponential Function**		61
	5.1	Calculus Definition	61
	5.2	Complex Exponentials	62
	5.3	Complex Trigonometry	63
	5.4	Matrix Exponentials	64
	5.5	Comments and Bibliography	67
	5.6	Exercises	67
		References	68
6	**ODE and Linear Algebra**		69
	6.1	Number Fields	69
	6.2	Vector Spaces	71
	6.3	Bases	74
	6.4	Linear Transformations	75
	6.5	Linear Equations	78
	6.6	Structure Theorem	81
	6.7	RC-Circuit Driven by a Simple Harmonic	83
	6.8	Principle of Superposition	86
	6.9	Comments and Bibliography	88
	6.10	Exercises	93
		References	95
7	**Linear ODE with Constant Coefficients**		97
	7.1	Eigendecomposition	98
	7.2	1DOF Mass-Spring System and Its Equivalent Circuit	102

	7.3	General Solutions of Homogeneous ODE	105
	7.4	Method of Undetermined Coefficients	106
	7.5	Convolution in One Dimension	109
	7.6	Multidimensional Convolution	116
	7.7	Impulse Response	117
	7.8	Step and Impulse Response of an RLC-Circuit	118
	7.9	Comments and Bibliography	121
	7.10	Exercises	123
		References	124
8	**Discrete Fourier Transform**		**125**
	8.1	Orthogonal Expansions in $\mathbb{R}^N$	125
	8.2	Real and Complex DFT	127
	8.3	Filtering Out Noise	132
	8.4	Comparison of Tides	135
	8.5	Forecasting Solar Activity	137
	8.6	DFT Solution of an ODE	138
	8.7	Comments and Bibliography	139
	8.8	Exercises	140
		References	142
9	**Fourier Series**		**143**
	9.1	Representation of Functions	144
	9.2	Inner Product Spaces	148
	9.3	Fourier Series Solutions of ODE	151
	9.4	RC-Circuit Driven by a Square Waveform	153
	9.5	Heat Equation on an Interval	159
	9.6	Comments and Bibliography	164
	9.7	Exercises	166
		References	168
10	**Fourier and Laplace Transforms**		**169**
	10.1	Fourier Transform	170
	10.2	Frequency Response Analysis	175
	10.3	Heat Equation on a Line	181
	10.4	Laplace Transform	182
	10.5	Comments and Bibliography	184
	10.6	Exercises	184
		References	185
Index			**187**

Example of Modeling Population Dynamics 1

The main subject of this introductory chapter is not mathematical biology, but the way ordinary differential equations (ODE) enter that subject and the manner in which they are used by demographers to analyse population data, such as the portion of the U.S. census record shown in Fig. 1.1.

Between the years 1910 and 2000 the size of the U.S. population had roughly tripled. Should similar growth be expected over the course of the 21st century?

1.1 Size-Limited Growth

Let P denote the U.S. population size. The plot in Fig. 1.1 suggests that P is a monotone increasing function of time t, yet that is all that can be said based on visual inspection. To find a formula for P, we must shift our attention to dP/dt: it is easier to make quantitative statements about the rate of change of a function than about the function itself.

The rate of growth of any population depends on its size. Assuming that size is the only limiting factor leads to the general growth model

$$\frac{dP}{dt} = f(P). \tag{1.1}$$

We have arrived at our first ODE. Equation (1.1) is *differential* because it involves the derivative of an unknown function P, and *ordinary* because the unknown function depends only on time; if P were multivariate, a relation between its partial derivatives would be called a partial differential equation (PDE). Equation (1.1) is also of *first order*, for that is the order of the highest derivative it contains, and *autonomous*, because its right-hand side does not

© The Author(s), under exclusive license to Springer Nature Switzerland AG 2025
A. Beltukov, *Differential Equations and Data Analysis*, Synthesis Lectures on
Mathematics & Statistics, https://doi.org/10.1007/978-3-031-62257-1_1

Year	Population
1910	92,228,496
1920	106,021,537
1930	123,202,624
1940	132,164,569
1950	151,325,798
1960	179,323,175
1970	203,211,926
1980	226,545,805
1990	248,709,873
2000	281,421,906

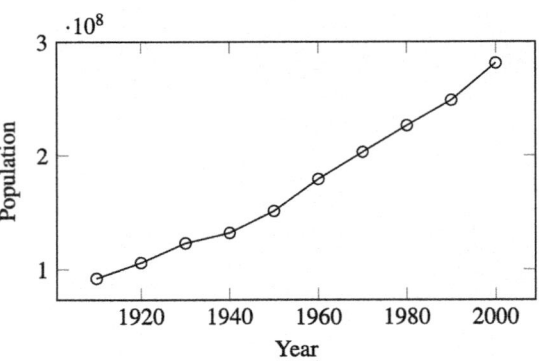

Fig. 1.1 Population of the United States in the 20th century (U.S. Census Bureau)

explicitly involve time; as such it can be solved using a Calculus technique called *separation of variables*.

Separating variables in (1.1) is premature—we should first make the right-hand side more explicit. To this end, let us expand f into a Taylor series at the origin and let us set the constant of the series to zero, so that in the absence of population there is no population growth:

$$f(P) = f'(0)\, P + \frac{1}{2} f''(0)\, P^2 + \frac{1}{6} f'''(0)\, P^3 + \ldots \tag{1.2}$$

If P is small, the series may be approximated with the sum of the first few terms. We will consider two scenarios in which (1.2) is truncated after the first and the second term, respectively.

1.2 Natural Growth Equation

Replacing the right-hand side of (1.1) with the first term of the series (1.2) gives the *natural growth equation*. We will write it as

$$\frac{dP}{dt} = a\, P \tag{1.3}$$

with $a = f'(0)$. The *growth constant* a has the units of inverse time and is positive, since the population is growing.

Being a linearization of (1.1), Eq. (1.3) is *linear*: more precisely, it is *linear homogeneous with constant coefficients*. The 'constant coefficients' part refers to the parameter a being a constant. As for the 'linear homogeneous' part, we have to postpone the explanation of what that means until the introduction of linear algebra language in Chap. 6. In the meantime, we will classify first order ODE as linear by isolating the derivative and checking whether it depends linearly on the function; if the derivative is directly proportional to the function, as is the case in (1.3), the ODE is linear homogeneous.

ODE (1.3) is simple enough for its solutions to be guessed. If we set $a = 1$ and translate the equation into common language, it says: the derivative of an unknown function is itself. This immediately brings to mind exponentials. Indeed, for $a = 1$ one obvious solution of (1.3) is $P = e^t$, but $P = 2e^t$ also works, as does $P = Ce^t$ where C is any constant. In general, the solutions of (1.3) are exponentials

$$P = Ce^{at}, \tag{1.4}$$

with C an arbitrary constant. Notice that, since a is measured in inverse time, the product at is non-dimensional, as it should be.

To validate (1.4), we can simply substitute it into (1.3):

$$\frac{dP}{dt} = \frac{d}{dt}\left(Ce^{at}\right) = a\left(Ce^{at}\right) = aP.$$

This shows that exponentials (1.4) satisfy ODE (1.3) and are therefore its solutions. The converse is also true but is less obvious: every solution of (1.3) is of the form (1.4). To prove this, write an arbitrary solution of (1.3) as $u(t)e^{at}$ where u is a function to be determined. Setting $P(t) = u(t)e^{at}$ in (1.3) gives

$$\frac{du}{dt}e^{at} + u(t)ae^{at} = au(t)e^{at},$$

which, after simplification, becomes $\frac{du}{dt} = 0$. This means that $u(t)$ is constant, so every solution of (1.3) is of the form (1.4).

Since Eq. (1.4) captures all solutions of ODE (1.3), it is the *general solution*. Having guessed it, we will now derive it constructively.

1.3 Separation of Variables

Suppose that, in an effort to recover P from dP/dt, we integrate both sides of (1.3) with respect to time:

$$P = a \int P(t)\,dt.$$

This leads to a dead end because the integral on the right requires the knowledge of how P depends on t, which is what we want to determine in the first place. To avoid $\int P(t)\,dt$, we must first separate P's from t's in (1.3) and then integrate, like so:

$$\int \frac{dP}{P} = \int a\,dt.$$

Now neither integral poses problems and carrying out indefinite integration gives

$$\ln(P) + C_1 = a\,t + C_2.$$

We dutifully added constants of integration on both sides but that is actually unnecessary: when we solve for P the two constants get consolidated into one

$$P = C\,e^{at}, \quad C = e^{C_2 - C_1}$$

and we get (1.4). Henceforth, when we integrate separable ODE, we will only add a single constant of integration to one side.

In Sect. 1.1 we mentioned that any first order autonomous ODE is separable. Indeed, we can separate variables in (1.1) as follows

$$\int \frac{dP}{f(P)} = \int dt.$$

Obviously, symbolic manipulation of the left-hand side integral requires specification of the function f. That prompted us to defer the discussion of separation of variables until the formulation of the natural growth equation.

More generally, one can separate variables in

$$\frac{dP}{dt} = f(t, P)$$

if the right-hand side can be factored into the product: $f(t, P) = f_1(t)\,f_2(P)$. Examples of separable ODE abound in calculus and ODE textbooks. We therefore omit them and return to population modeling.

1.4 Use of Initial Values

In order to match the general solution (1.4) of the natural growth equation against the census data in Fig. 1.1, we need numerical values of a and C. These parameters are different in nature: the growth constant a is intrinsic to the model, whereas C entered the solution through indefinite integration. We will estimate a in the next section. Here we will determine C from the *initial value* $P(t_0) = P_0$: for the data in Fig. 1.1, $t_0 = 1910$ and $P_0 = P(1910) = 92{,}228{,}496$.

Evaluating both sides of Eq. (1.4) at $t = t_0$ gives $P(t_0) = P_0 = C\,e^{a\,t_0}$. It follows that $C = P_0\,e^{-a\,t_0}$ and

$$P = P_0\,e^{a\,(t - t_0)}. \tag{1.5}$$

Equation (1.5) is the solution of the *initial value problem* (IVP) consisting of ODE (1.3) and the initial condition $P(t_0) = P_0$. Almost all differential equations considered in the remainder of the book will come as initial value problems.

1.5 Exponential Fit via Linear Regression

Estimation of model parameters generally requires nonlinear regression which is the subject of Chap. 2. However, Eq. (1.5) can be linearized using semi-logarithmic coordinates which allows the use of simpler linear regression.

It will be convenient to introduce vector notation. Let **t** and **P** be the first and second columns of the table in Fig. 1.1, regarded as column vectors in $\mathbb{R}^{10}$. Also, let $\mathbf{x} = \mathbf{t} - t_0$ and $\mathbf{y} = \ln(\mathbf{P}/P_0)$, where the subtraction of the scalar and the application of the logarithm are componentwise—henceforth this is how all functions will be applied to vectors. As follows from Eq. (1.5), $a\,\mathbf{x} = \mathbf{y}$, so, theoretically, a can be found by dividing any component of **y** by the corresponding component of **x**, as long as the latter is nonzero. However, due to noise, it is certain that **x** and **y** are misaligned and $a\,\mathbf{x}$ cannot equal **y** for any value of a. Since solving $a\,\mathbf{x} = \mathbf{y}$ for a is impossible, the best we can do is minimize the magnitude of the difference $a\,\mathbf{x} - \mathbf{y}$; we will call that difference the *residual* and refer to its magnitude $\|a\,\mathbf{x} - \mathbf{y}\|$ as the 2-*norm*.

Minimizing the norm is equivalent to minimizing its square, and the square of the 2-norm is the dot product of the vector with itself:

$$\|a\,\mathbf{x} - \mathbf{y}\|^2 = (a\,\mathbf{x} - \mathbf{y}) \cdot (a\,\mathbf{x} - \mathbf{y}) = a^2\,(\mathbf{x} \cdot \mathbf{x}) - 2\,a\,(\mathbf{x} \cdot \mathbf{y}) + (\mathbf{y} \cdot \mathbf{y}).$$

Evidently, $\|a\,\mathbf{x} - \mathbf{y}\|^2$ is a quadratic in a and so the minimum is at

$$a = \frac{\mathbf{x} \cdot \mathbf{y}}{\mathbf{x} \cdot \mathbf{x}}. \tag{1.6}$$

Application of Eq. (1.6) to the data in Fig. 1.1 gives $a \approx 0.0126\ \text{Year}^{-1}$ and the fit shown in Fig. 1.2.

To test the predictive power of (1.5), we compared it to the two most recent census results (filled circles). For 2010 the model is spot on, giving relative error of less than 0.1%, but for 2020 the relative error is more than 12%.

On one hand, Fig. 1.2 shows that exponential growth is appropriate for coarsely describing the dynamics of the U.S. population in the 20th century. On the other hand, the large overestimate for 2020 casts doubt on whether that growth can be sustained throughout the 21st century. This suggests replacing the natural growth equation with a more sophisticated model.

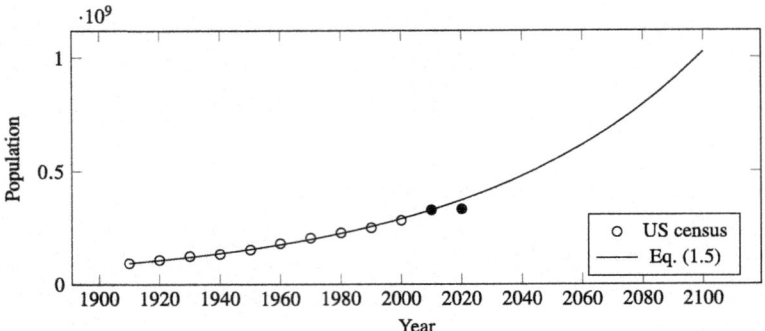

Fig. 1.2 Exponential model (1.5) fitted to the U.S. census data in Fig. 1.1 (empty circles); the filled circles are the two most recent census results

1.6 Logistic Equation

The logistic equation is obtained by approximating the right-hand side of (1.1) with the first two terms of the Taylor series (1.2). We will write it as

$$\frac{dP}{dt} = aP - bP^2, \tag{1.7}$$

where $a = f'(0) > 0$ and $b = -\frac{1}{2} f''(0) > 0$; the minus sign in front of b ensures that the quadratic term impedes population growth.

Being first order and autonomous, the logistic Eq. (1.7) is separable:

$$\int \frac{dP}{aP - bP^2} = \int dt.$$

The integral on the left requires partial fraction decomposition

$$\int \frac{dP}{aP - bP^2} = \frac{1}{a} \left(\int \frac{dP}{P} + \int \frac{b\,dP}{a - bP} \right) = \frac{1}{a} \ln(P) - \frac{1}{a} \ln(a - bP).$$

We omitted the constant of integration because it will be added on the right hand side: $\int dt = t + C$. Combining the logarithms and solving

$$\frac{1}{a} \ln \frac{P}{a - bP} = t + C$$

for P yields

$$P = \frac{a}{b + C_1 e^{-at}}, \quad C_1 = e^{-aC}.$$

Finally, imposing the same initial condition as in Sect. 1.4 gives

1.6 Logistic Equation

$$P = \frac{a}{b + \left(\frac{a}{P_0} - b\right) e^{-a(t-t_0)}}. \tag{1.8}$$

We leave it as an exercise to symbolically confirm that (1.8) satisfies ODE (1.7) and the initial condition $P(t_0) = P_0$. Instead, we will validate (1.8) numerically in MATLAB.

The following code solves (1.7) using ode45 and compares the numerical solution with (1.8).

```
a = 1; b = .2; c = a/b;
t0 = 3; P0 = .05;
odefun = @(t,P) a*P-b*P^2;
[t,P1] = ode45(odefun,[t0 t0+10],P0);
P2 = c./(1+(c/P0-1)*exp(-a*(t-t0)));
plot(t,P1,'ko',t,P2,'k-');
legend('ode45','exact','location','east');
xlabel('Time'); ylabel('Population');
title(sprintf('Max error %1.2e',norm(P1-P2,inf)));
```

Figure 1.3 shows that the numerical solution matches the symbolic solution; the maximum error of 4.71×10^{-4} (computed as ∞-norm) is consistent with default error tolerances for ode45. The reader can check that the circles stay on the line if the parameters a, b, and the initial condition are changed.

Figure 1.3 further shows that logistically growing population stabilizes at the level

$$\lim_{t \to \infty} P(t) = \frac{a}{b}.$$

This limiting value is called *carrying capacity* and denoted P_∞.

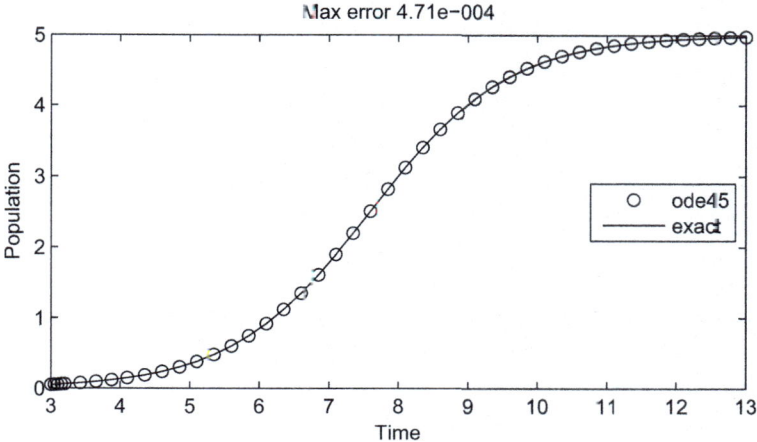

Fig. 1.3 Validation of Eq. (1.8) with $a = 1$, $b = 0.2$, $P(3) = 0.05$

1.7 Logistic Fit via the Simpex Method

Fitting the logistic curve (1.8) directly to the data in Fig. 1.1 is complicated by the fact that the parameters a and b differ by 9 orders of magnitude: $a \sim 10^{-2}$ Year^{-1}, whereas $b \sim 10^{-11}$ Year^{-1}. For this reason we will perform the logistic fit as a special type of exponential fit.

As in Sect. 1.5, let $\mathbf{t}$ and $\mathbf{P}$ be the columns of the table in Fig. 1.1. Also, let $\mathbf{x} = \mathbf{t} - t_0$ and $\mathbf{y} = P_0/\mathbf{P}$; recall that our convention is to apply functions to vectors componentwise. As follows from Eq. (1.8), the relationship between $\mathbf{x}$ and $\mathbf{y}$ is exponential

$$\mathbf{y} = c + (1-c)\, e^{-a\mathbf{x}}, \quad c = \frac{P_0}{(a/b)} = \frac{P_0}{P_\infty}. \tag{1.9}$$

Crucially, the non-dimensional parameter c, which happens to be the ratio of the initial population to carrying capacity, has the same order of magnitude as a.

In principle, a and c can be determined from three data points. To simplify notation, let us index the points (x_n, y_n) using $n = 1, 2, 3$ with the understanding that these are any three distinct points from the data set. Elimination of c from the system

$$c + (1-c)\, e^{-a x_n} = y_n, \quad n = 1, 2, 3 \tag{1.10}$$

leads to

$$\frac{e^{-a x_2} - e^{-a x_1}}{e^{-a x_3} - e^{-a x_2}} = \frac{y_2 - y_1}{y_3 - y_2}.$$

If x_n's are chosen so that $x_3 - x_1 = 2(x_2 - x_1) = 2T$ then

$$\frac{e^{-a x_2} - e^{-a x_1}}{e^{-a x_3} - e^{-a x_2}} = \frac{e^{-a(x_2 - x_1)} - 1}{e^{-a(x_3 - x_1)} - e^{-a(x_2 - x_1)}} = \frac{e^{-a T} - 1}{e^{-a 2 T} - e^{-a T}} = e^{a T}.$$

Consequently,

$$a = \frac{1}{T} \ln \frac{y_2 - y_1}{y_3 - y_2}. \tag{1.11}$$

Once a is determined using (1.11), the parameter c can be found from (1.10). For instance, if $x_2 \neq 0$ then

$$c = \frac{y_2 - e^{-a x_2}}{1 - e^{-a x_2}}. \tag{1.12}$$

In practice, it is common to set x_1 and x_3 to the endpoints of the x-interval and use the midpoint as x_2. If the value y_2 corresponding to the midpoint x_2 is absent in the data set, it can be interpolated.

Of course, due to noise, the values given by Eqs. (1.11) and (1.12) are unlikely to be accurate. However, they can be used as a starting point for an iterative optimization routine.

The following code minimizes $\|c + (1-c)\, e^{-a\mathbf{x}} - \mathbf{y}\|$ with respect to c and a, using the starting values given by (1.11) and (1.12). Proper minimization requires setting the gradient

1.7 Logistic Fit via the Simpex Method

to zero and solving the resulting equations using multidimensional Newton's method—we postpone that until Chap. 2. In the meantime we use fminsearch, which is MATLAB's implementation of the Nelder-Mead simplex method.

```
x  = t - t(1);      y  = P(1)./P;
x1 = x(1);   x3 = x(end);   x2 = .5*(x1 + x3);
y1 = y(1);   y3 = y(end);   y2 = interp1(x,y,x2);
T  = .5*(x3 - x1);
a  = log((y2 - y1)/(y3-y2))/T;
c  = (y2 - exp(-a*x2))/(1 - exp(-a*x2));
fun = @(z) norm(z(1) + (1 - z(1))*exp(-z(2)*x) - y);
z  = fminsearch(fun, [c;a]);
```

According to fminsearch, $a \approx 0.01430$ Year^{-1} and $c \approx 0.05903$ which gives the fit shown in Fig. 1.4.

The logistic model projects the US population in 2100 to be 706,525,832, about 53% of the carrying capacity $P_\infty = 1,336,138,120$. For the data in Fig. 1.1 (empty circles) the 3.3% maximum relative error of the logistic model is comparable to the 2.5% maximum relative error of the exponential model. However, unlike the exponential model, which gives a very accurate projection for 2010 and a very inaccurate projection for 2020, the logistic model chooses the middle ground, with relative error in the range of 4% to 5% in each case. Most importantly, the logistic model predicts equilibrium (achieved sometime in mid-25th century) which is more realistic behavior than unbounded exponential growth.

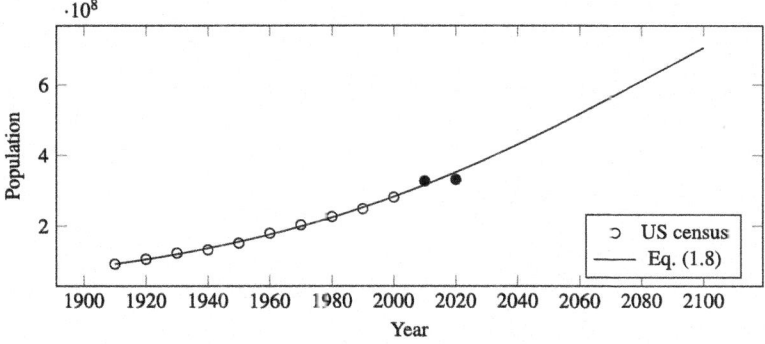

Fig. 1.4 Logistic model (1.8) fitted to the U.S. census data in Fig. 1.1 (empty circles); the filled circles are the two most recent census results

1.8 Comments and Bibliography

The reasoning in Sect. 1.1 can be found in Lotka's classic "Elements of Mathematical Biology" [3]. We highly recommend Lotka's book both for its style and the wealth of included experimental data.

In "An Essay on the Principle of Population" (published in 1798) Thomas Malthus postulated that human population, when unchecked, increases in a geometrical ratio (exponentially) while subsistence increases only in an arithmetical ratio (linearly); the natural growth Eq. (1.3) is often traced to that essay. Malthus predicted that poverty, hunger, and armed conflicts were unavoidable, since sooner or later humanity must outgrow available resources: this is known as *Malthusian catastrophe*.

The logistic Eq. (1.7) was published by Pierre-François Verhulst in 1845. Verhulst's starting point was the natural growth equation in which he replaced the growth constant with a linear function of population size. For a complete historical account of the development of mathematical population dynamics, consult [1] and the references therein.

There is a striking similarity between the logistic model and the kinetics of the bimolecular reaction

$$X + Y \xrightarrow{k} Z$$

Let x, y, and z be the concentrations of species X, Y and Z, respectively. According to the *law of mass action* (formulated by Guldberg and Waage between 1864 and 1879)

$$\frac{dx}{dt} = -k\,x\,y, \quad \frac{dy}{dt} = -k\,x\,y, \quad \frac{dz}{dt} = k\,x\,y,$$

where k is a positive rate constant. From the equality of the rates $dx/dt = dy/dt$ follows mass balance $x - x_0 = y - y_0$. Therefore $y = x - x_0 + y_0$ and, consequently,

$$\frac{dx}{dt} = -k\,x\,(x - x_0 + y_0) = k\,(x_0 - y_0)\,x - k\,x^2.$$

This is the logistic equation (1.7) in disguise. The parameter $a = k\,(x_0 - y_0)$ may not be positive if $y_0 \geq x_0$. Still a valid solution can be obtained from Eq. (1.8) by substituting $P = x$, $a = k\,(x_0 - y_0)$ and $b = k$.

Natural growth also has its counterpart in chemical kinetics. For the monomolecular reaction

$$X \xrightarrow{k} Y$$

the law of mass action gives the system

$$\frac{dx}{dt} = -k\,x, \quad \frac{dy}{dt} = k\,x. \tag{1.13}$$

1.8 Comments and Bibliography

The ODE for x is identical in form to (1.3) but the parameter $a = -k$ is negative, so instead of exponential growth there is exponential decay.

The similarity between bimolecular kinetics and the logistic equation was noted around 1908 by the American physiologist T. Brailsford Robertson and, independently, by the German chemist Wolfgang Ostwald. The connection is not accidental. In bimolecular reactions chemical bonds are formed during random molecular collisions. Meanwhile, certain biological phenomena, such as predation and sexual reproduction, may also be regarded as random collisions, if the populations are large enough. The similarity between natural growth and (1.13) is also not accidental as will be explained in Chap. 4.

One can think of the natural growth equation and the logistic equation as the beginning of an infinite progression of models obtained by retaining more and more terms in the series (1.2). Each term adds a parameter which can be used to improve agreement between the model and the data. Unfortunately, there is a catch: the more parameters a mathematical model has, the weaker is its predictive power. As John von Neumann famously quipped: "With four parameters I can fit an elephant, and with five I can make him wiggle his trunk."

The proof that (1.4) is the general solution of (1.3) can be found in [2]; it foreshadows a useful technique for solving linear nonhomogeneous ODE called *variation of parameters* which is featured in Chap. 7.

In Sect. 1.5 we introduced 'norm' as a synonym of 'magnitude' but the two words are not merely synonyms. 'Magnitude' is a nebulous common word, whereas 'norm' in mathematics has the precise meaning of a function that assigns nonnegative numbers to vectors subject to certain axioms. We should also mention that 2-norm is a member of an important family of p-norms defined by

$$\|\mathbf{x}\|_p = \left(\sum_{n=1}^{N} |x_n|^p\right)^{\frac{1}{p}}, \quad p \geq 1.$$

By default, we will use the 2-norm and omit the subscript. However, on occasion we will use the 1-norm—the sum of absolute values—and the infinity norm:

$$\|\mathbf{x}\|_\infty = \lim_{p \to \infty} \|\mathbf{x}\|_p = \max\{|x_1|, \ldots, |x_N|\}.$$

In fact, we have already used the infinity norm in Sect. 1.6 to compare the output of ode45 to the symbolic solution (1.8) of the logistic equation.

Like any nonlinear fit, the logistic fit in Fig. 1.4 is very sensitive to changes in the data. If we augment the table in Fig. 1.1 with the census count for 2010, the estimate of the carrying capacity jumps from little over 1.3 billion to nearly 1.8 billion, yet, if we further add the data for 2020, the carrying capacity drops to about 1.1 billion: this is one reason why Fig. 1.4 should be taken with a grain of salt. Other reasons include age distribution, immigration, technological progress, climate change, and a myriad of other factors that the logistic model does not take into account. The logistic model is reasonably accurate only for simple organisms growing in steady conditions, like yeast in a Petri dish. Still, it is used by demographers for short-term forecasts and planning.

1.9 Exercises

1. Consider the following IVP

$$\frac{dP}{dt} = aP + b, \quad P(0) = P_0, \tag{1.14}$$

   where a and b are positive; this could be natural growth with constant rate of immigration. Solve (1.14) and validate the solution numerically, using `ode45`.

2. Consider the following model of population growth

$$\frac{dP}{dt} = aP^2, \quad P(0) = P_0, \tag{1.15}$$

   where a is positive. Show that population becomes infinite in finite time—this is worse than Malthusian catastrophe.

3. Solve the following variant of the logistic equation (using separation of variables and partial fraction decomposition)

$$\frac{dP}{dt} = aP - bP^3, \quad P(0) = P_0, \tag{1.16}$$

   and fit the solution to the data in Fig. 1.1 using `fminsearch`. Compare the accuracy of the model (1.16) with that of the logistic equation (1.7).

4. This problem requires some knowledge of probability and statistics. Consider IVP (1.14) with the initial condition $P(0)$ that is a normally distributed random variable with mean μ_0 and variance σ_0^2. Show that the solution $P(t)$ is also a normally distributed random variable and find its mean μ_t and variance σ_t^2.

5. Suppose that $f(x, y)$ is twice-differentiable. Construct an expression

$$u\left(f, \frac{\partial f}{\partial x}, \frac{\partial f}{\partial y}, \frac{\partial^2 f}{\partial x^2}, \frac{\partial^2 f}{\partial x \partial y}, \frac{\partial^2 f}{\partial y^2}\right)$$

   that is identically zero if and only if $f(x, y) = f_1(x) f_2(y)$. Put differently, find a partial differential equation (PDE) whose solutions are separable functions of two variables. The answer can be found in [4].

References

1. N. Bacaër, *A Short History of Mathematical Population Dynamics* (Springer, London, 2011)
2. MW. Hirsch, S. Smale, *Differential Equations, Dynamical Systems, and Linear Algebra*. Number 60 in Pure and applied mathematics (Academic Press, 1974)
3. A.J. Lotka, *Elements of Mathematical Biology* (Dover books on biology, Dover Publications, 1956)
4. David Scott, When is an ordinary differential equation separable? Am. Math. Mon. **92**(6), 422–423 (1985)

Estimation of Parameters

2

Once an ODE model is formulated and solved, its parameters must be determined from experimental data. This can be the most challenging part of the modeling process.

Let $y = f(x, p_1, \ldots, p_M)$ be the solution of a model that relates scalar variables x and y and contains parameters $p_1, \ldots, p_M$. The standard procedure for estimating these parameters from data—*nonlinear least squares* (NLS)—is to minimize the square of the 2-norm of the residual

$$g(p_1, \ldots, p_M) = \sum_{j=1}^{N} (f(x_j, p_1, \ldots, p_M) - y_j)^2. \tag{2.1}$$

As we show in Sect. 2.1, NLS produces *maximum likelihood estimates* (MLE): this is what makes it standard. On occasion, when the data has visible outliers, the 2-norm in (2.1) may be replaced with 1-norm which, as we show in Sect. 2.6, is more robust. However such occasions are rare and we will mostly use NLS.

If the function (2.1) is differentiable, its minimum can be found by setting the gradient to zero

$$\frac{\partial g}{\partial p_1} = 0, \ \ldots, \ \frac{\partial g}{\partial p_M} = 0. \tag{2.2}$$

Systems of nonlinear algebraic equations are commonly solved using *Newton's method* which is the subject of Sects. 2.2 and 2.4.

Section 2.1 on MLE requires some familiarity with the calculus of random variables and the central limit theorem; we hope that the reader who lacks that familiarity will be inspired to acquire it. Explanation of multidimensional Newton's method in Sect. 2.4 uses some notions from linear algebra that are reinforced in Chap. 6.

2.1 Maximum Likelihood Estimation

Roughly speaking, *likelihood* is the probability of observing experimental data given the values of the underlying model's parameters. As a simple example, consider an experiment consisting of 10 coin tosses where 7 outcomes are heads. Let p be the mathematical probability of heads

$$p = \lim_{N \to \infty} \frac{\text{\# of heads out of } N \text{ tosses}}{N}.$$

For 10 tosses the probability of observing 7 heads is given by the binomial distribution with parameter p; when considered as a function of that parameter, it becomes the likelihood

$$L(p) = \binom{10}{7} p^7 (1-p)^3. \tag{2.3}$$

Figure 2.1 shows that (2.3) has a global maximum at $\widehat{p} = .7$: this is the maximum likelihood estimate of p. In the absence of other information about the coin, $\widehat{p}$ is the best estimate of the mathematical probability of heads.

As a more subtle example, consider the problem of estimating parameters of the normal distribution $N(\mu, \sigma)$ from two sample values

$$y_1 = 0.537667139546100, \quad y_2 = 1.833885014595087.$$

Here the likelihood cannot be set to mathematical probability of observing these exact numbers because that probability is zero. Yet the numbers are not exact: they were generated on the computer with MATLAB's normal random number generator randn and should thus be regarded as true values truncated after 15 decimals. Put differently, the sample data is specified with *double precision* $\pm \epsilon$ where $\epsilon \approx 2 \times 10^{-16}$ is the *machine epsilon*—the smallest positive number that the computer can distinguish from zero. Due to finite precision,

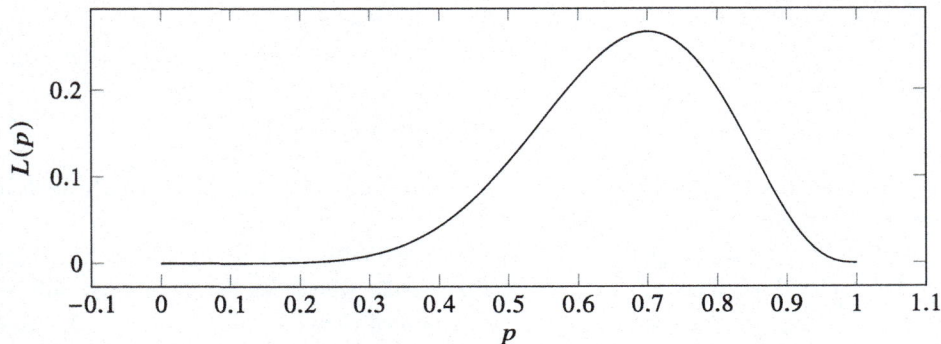

Fig. 2.1 Likelihood function for a binomial experiment with $n = 10$ and $k = 7$

2.1 Maximum Likelihood Estimation

the probability of observing $\{y_1, y_2\}$ is positive, and is that of the true sample values (with infinitely many decimals) landing inside the box with center (y_1, y_2) and sides of length 2ϵ:

$$\int_{y_2-\epsilon}^{y_2+\epsilon} \int_{y_1-\epsilon}^{y_1+\epsilon} \frac{1}{\sqrt{2\pi\sigma^2}} e^{-\frac{(x_1-\mu)^2}{2\sigma^2}} \frac{1}{\sqrt{2\pi\sigma^2}} e^{-\frac{(x_2-\mu)^2}{2\sigma^2}} dx_1 dx_2.$$

We can define the likelihood of $\{y_1, y_2\}$ as that probability. However, since ϵ is small, we can approximate the integral as

$$(2\epsilon)^2 \frac{1}{\sqrt{2\pi\sigma^2}} e^{-\frac{(y_1-\mu)^2}{2\sigma^2}} \frac{1}{\sqrt{2\pi\sigma^2}} e^{-\frac{(y_2-\mu)^2}{2\sigma^2}}$$

which is simpler to maximize. The factor $(2\epsilon)^2$ does not affect the location of the maximum and may be dropped. The likelihood of $\{y_1, y_2\}$ is then simply the joint probability density of $\{y_1, y_2\}$ regarded as a function of μ and σ:

$$L(\mu, \sigma) = \left(\frac{1}{\sqrt{2\pi\sigma^2}}\right)^2 \exp\left(-\frac{1}{2\sigma^2} \sum_{j=1}^{2}(y_j - \mu)^2\right) \quad (2.4)$$

Setting the gradient of (2.4) to zero and solving the resulting equations gives

$$\widehat{\mu} = \frac{1}{2}\sum_{j=1}^{2} y_j = 1.185776077070593$$

$$\widehat{\sigma}^2 = \frac{1}{2}\sum_{j=1}^{2} (y_j - \widehat{\mu})^2 = 0.420045194899127.$$

These maximum likelihood estimates are far from the parameters of the standard normal distribution $N(0, 1)$ that was used to generate the data. Yet that is to be expected when the sample is so small.

We are now ready to consider the main problem of this chapter: estimating parameters $p_1, \ldots, p_M$ from the sample of values of $y = f(x, p_1, \ldots, p_M)$ contaminated by noise. Supposing the errors in the data to be sums of numerous independent errors, we can use the central limit theorem to argue that they are normally distributed. In other words, we may assume that

$$y_j = f(x_j, p_1, \ldots, p_M) + \epsilon_j, \quad j = 1, \ldots, N,$$

where ϵ_j's are independent normal variables with zero means and common variance σ^2— this is the definition of *white noise*. The data values are then independent normal variables

$$y_j \sim N(f(x_j, p_1, \ldots, p_M), \sigma^2)$$

with the joint probability density function—the likelihood of $\{(x_j, y_j)\}_{j=1}^N$—being the product of Gaussians. That product can be combined into a single Gaussian with (2.1) inside the exponential:

$$L(p_1, \ldots, p_M, \sigma) = \prod_{j=1}^N \frac{1}{\sqrt{2\pi\sigma^2}} \exp\left(-\frac{(y_j - f(x_j, p_1, \ldots, p_M))^2}{2\sigma^2}\right)$$
$$= (2\pi)^{-\frac{N}{2}} \sigma^{-N} \exp\left(-\frac{g(p_1, \ldots, p_M)}{2\sigma^2}\right). \tag{2.5}$$

Setting the partial derivatives of (2.5) with respect to p-variables to zero implies Eq. (2.2): thus nonlinear least squares estimates of $p_1, \ldots, p_M$ are the same as the maximum likelihood estimates $\widehat{p}_1, \ldots, \widehat{p}_M$. Since (2.5) also depends on σ, we must set $\frac{\partial L}{\partial \sigma} = 0$. This leads to MLE of variance

$$\widehat{\sigma}^2 = \frac{1}{N} \sum_{j=1}^N (f(x_j, \widehat{p}_1, \ldots, \widehat{p}_M) - y_j)^2. \tag{2.6}$$

Having justified Eq. (2.2), let us turn our attention to its solution. We will first consider the simplest case when there is only one parameter to be estimated.

2.2 Newton's Method in One Dimension

Let p_* be a root of a nonlinear equation $h(p) = 0$. Newton's method is the iteration

$$p_{n+1} = p_n - \frac{h(p_n)}{h'(p_n)}. \tag{2.7}$$

The sequence $\{p_n\}$ often (but not always!) converges to p_* in the manner illustrated in Fig. 2.2.

n	p_n
1	10.000000000000000
2	5.150000000000000
3	2.866262135922330
4	1.956460731776899
5	1.744920939145020
6	1.732098271119538
7	1.732050808219183
8	1.732050807568877

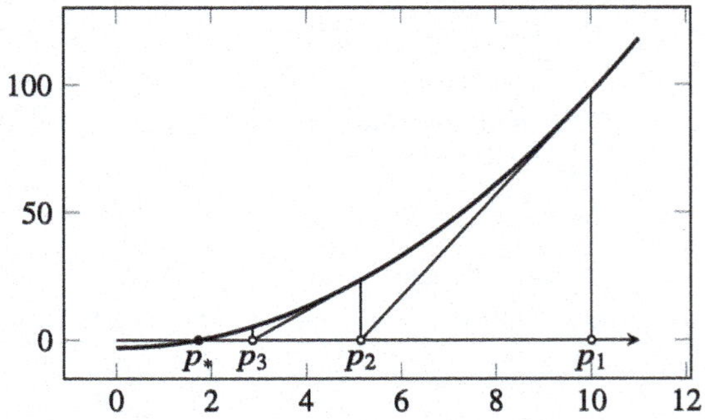

Fig. 2.2 Newton's method applied to $h(p) = p^2 - 3 = 0$ with the starting value $p_1 = 10$; the zigzag path is made of tangents to $h = h(p)$

2.2 Newton's Method in One Dimension

The idea behind Eq. (2.7) is that near a simple root the function is close to its tangent line. That is, if p_n is close to p_* then linearizing h at p_n and solving

$$h(p_n) + h'(p_n)(p - p_n) = 0$$

generally (but, again, not always) gives a better approximation p_{n+1}.

In Fig. 2.2 Newton's method is applied to the quadratic equation $p^2 - 3 = 0$ for which Newton's iteration (2.7) is

$$p_{n+1} = \frac{1}{2}\left(p_n + \frac{3}{p_n}\right).$$

Even with a poor starting value $p_1 = 10$, Newton's method takes only 7 iterations to yield $\sqrt{3}$ with 15 digits of accuracy. Remarkably, this method of computation of square roots was known to Heron of Alexandria!

For a quadratic equation Newton's method converges to the root closest to the starting value, which may be complex; if the starting value is equidistant from the roots, the method does not converge. The same cannot be said about more complicated equations: for some starting values Newton's method may diverge to infinity, while for others it may get stuck in a cycle, like the one shown in Fig. 2.3.

To produce Fig. 2.3 we applied Newton's method to $p^3 - 2p = 0$ with a carefully chosen starting value $p_1 = \frac{\sqrt{10}}{5}$. In exact arithmetics, the resulting sequence should be

$$\frac{\sqrt{10}}{5}, -\frac{\sqrt{10}}{5}, \frac{\sqrt{10}}{5}, -\frac{\sqrt{10}}{5}, \ldots$$

However, due to roundoff errors, Newton's method breaks out of the cycle and converges to $-\sqrt{2}$ after 31 iterations.

The cycle in Fig. 2.3 may look uncomplicated and, thanks to roundoff, harmless. However, let us look at the behavior of Newton's iteration for $p^3 - 2p = 0$ when the starting values are points in the complex plane.

n	p_n
1	0.632455532033676
2	−0.632455532033676
3	0.632455532033677
4	−0.632455532033684
5	0.632455532033724
6	−0.632455532033964
7	0.632455532035402
8	−0.632455532044034

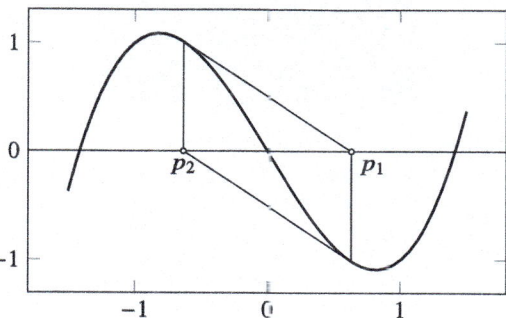

Fig. 2.3 Newton's metd applied to $p^3 - 2p = 0$ with starting value $p_1 = \frac{\sqrt{10}}{5}$

The cubic $p^3 - 2p = 0$ has three roots: $-\sqrt{2}, 0,$ or $\sqrt{2}$. For each complex starting value, the iteration

$$p_{n+1} = p_n - \frac{p_n^3 - 2 p_n}{3 p_n^2 - 2}$$

will converge to one of these roots. If we color the points of the complex plane based on that principle, the result is the map of *basins of attraction* of the roots whose portion is shown in Fig. 2.4; color gradation represents the number of iterations required for convergence.

The boundaries of the basins of attraction in Fig. 2.4 are *fractals*: zooming in on any of the teardrop features, brings out more and more such features, ad infinitum.

Any point in the interior of a basin of attraction can be surrounded with a disk so that for all starting values within that disk Newton's method converges to one and the same root—the root for that basin. In contrast, arbitrarily small neighborhoods of a point on the boundary of a basin of attraction contain starting values for all roots. This makes basins of attraction visually stunning and endlessly fascinating but, on the downside, this also means that small perturbation of the starting value may nudge Newton's method away from the root we seek to a completely different root.

While Newton's method can be capricious at times, it has the redeeming feature of being *quadratically convergent*: after the onset of convergence, the number of significant digits doubles with every iteration. This can be seen in the iterates shown in Fig. 2.2 and is even more conspicuous in the next example.

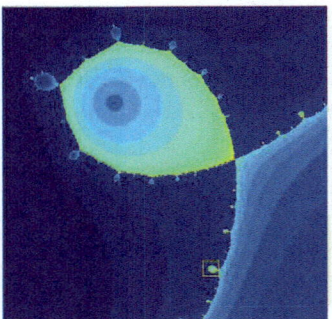 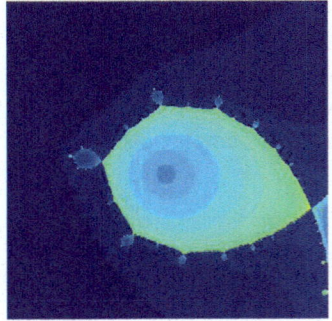

Fig. 2.4 Basins of attraction for the roots of $p^3 - 2p$; the lightest color corresponds to $-\sqrt{2}$. The image on the left was produced by taking starting values $x + yi$ with $6.5 \leq x \leq 8.5$ and $10 \leq y \leq 12$; the image on the right is the magnified detail of the left image marked with a square

2.3 One-Parameter Exponential Fit

In Sect. 1.5 we linearized the U.S. census data in Fig. 1.1 so that the growth constant a in Eq. (1.5) could be estimated through linear regression. We will now find a better value for that constant using nonlinear regression.

As in Sect. 1.5, we let $\mathbf{t}$ and $\mathbf{P}$ be the columns of data in Fig. 1.1; we also set $\mathbf{x} = \mathbf{t} - t_0$, $\mathbf{y} = \ln(\mathbf{P}/P_0)$, and use the estimate (1.5) as the starting value:

$$a_1 = \frac{\mathbf{x} \cdot \mathbf{y}}{\mathbf{x} \cdot \mathbf{x}}.$$

NLS estimate $\hat{a}$ is the minimum of $\|P_0 \, e^{a \, (\mathbf{t}-t_0)} - \mathbf{P}\|^2$ which is the same as the minimum of $g(a) = \|e^{a\mathbf{x}} - \mathbf{z}\|^2$ with $\mathbf{z} = \mathbf{P}/P_0$. Accordingly, we will search for the root of

$$h(a) = \frac{dg}{da} = \frac{d}{da}\left(e^{a\mathbf{x}} - \mathbf{z}\right) \cdot \left(e^{a\mathbf{x}} - \mathbf{z}\right) = 2\,\mathbf{x}\,e^{a\mathbf{x}} \cdot \left(e^{a\mathbf{x}} - \mathbf{z}\right).$$

The derivative of h can be simplified using the symmetry of the dot product:

$$\begin{aligned}\frac{dh}{da} &= 2\,\mathbf{x}^2\,e^{a\mathbf{x}} \cdot \left(e^{a\mathbf{x}} - \mathbf{z}\right) + 2\,\mathbf{x}\,e^{a\mathbf{x}} \cdot \mathbf{x}\,e^{a\mathbf{x}} \\ &= 2\,\mathbf{x}\,e^{a\mathbf{x}} \cdot \left(2\,\mathbf{x}\,e^{a\mathbf{x}} - \mathbf{x}\,\mathbf{z}\right).\end{aligned}$$

Newton's iterations (2.7) are then

$$a_{n+1} = a_n - \frac{\mathbf{x}\,e^{a_n \mathbf{x}} \cdot (e^{a_n \mathbf{x}} - \mathbf{z})}{\mathbf{x}\,e^{a_n \mathbf{x}} \cdot (2\,\mathbf{x}\,e^{a_n \mathbf{x}} - \mathbf{x}\,\mathbf{z})}, \qquad (2.8)$$

where we need to update only $e^{a\mathbf{x}}$ and $\mathbf{x}\,e^{a\mathbf{x}}$, as we do in the code below; before running the code, the variables t and P should be initialized as columns of the data from Fig. 1.1.

```
x = t - t(1); z = P/P(1); y = log(z); xz = x.*z;
a1 = dot(x,y)/dot(x,x);
for n=1:20
    u = exp(a1*x); ux = u.*x;
    a2 = a1 - dot(ux,u-z)/dot(ux,2*ux-xz);
    if abs(a2-a1) < eps('double')
        break;
    else
        a1 = a2;
    end
end
```

In a clear display of quadratic convergence, Newton's method stops after 4 iterations giving the NLS/MLE estimate

n	a_n
1	0.012670224988108
2	0.012582135882014
3	0.012581221733536
4	0.012581221636152

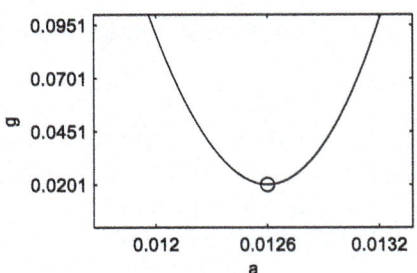

Fig. 2.5 Estimation of the growth constant a in Eq. (1.5) using the data in Fig. 1.1. The left panel shows Newton's iterations with starting value given by (1.6); the plot on the right confirms the minimum

$$\widehat{a} = 0.012581221636152 \text{ Year}^{-1}.$$

The corresponding estimate of the standard deviation (2.6) can be computed as follows:

$$\widehat{\sigma} = \frac{1}{\sqrt{10}} P_0 \| e^{\widehat{a}\mathbf{x}} - \mathbf{z} \| = 0.044\,859\,214\,499\,361\, P_0.$$

The standard deviation is thus about 4.5% of the initial population size.

Figure 2.5 shows the sequence a_n and the plot of $g(a) = \|e^{a\mathbf{x}} - \mathbf{z}\|^2$ which confirms that $\widehat{a}$ is indeed a minimum.

Since $\widehat{a}$ is close to the starting value produced with (1.6), an argument can be made that, in this particular case, nonlinear least squares method is not worth the bother. If we had to perform computations by hand then, certainly, we would have limited ourselves to using (1.6). However, it does not take much effort for a computer to refine the simple estimate (1.6) to MLE. Therefore we see no reason to avoid nonlinear least squares.

2.4 Newton's Method for Systems

Formulation of Newton's method for systems requires some familiarity with matrix algebra. We assume that the reader is familiar with Gaussian elimination (backslash in MATLAB), the row-by-column rule of matrix multiplication, and matrix inversion. Most importantly, we expect the reader to know how to write quadratic Taylor approximation of a multivariate function using gradient and Hessian.

A system of M scalar equations in M scalar unknowns

$$h_1(p_1, \ldots, p_M) = 0, \ldots, h_M(p_1, \ldots, p_M) = 0$$

can be aggregated into a single vector equation with a single vector unknown, which we will write compactly as $\mathbf{h}(\mathbf{p}) = \mathbf{0}$. Now that we have a single equation with a single unknown, we can arrive at Newton's method by following the same line of reasoning as in Sect. 2.2.

2.5 Two-Parameter Exponential Fit

Recall that the idea for solving $h(p) = 0$ is to improve an approximation to the root by solving the linearized equation. Linearization of $\mathbf{h}(\mathbf{p}) = \mathbf{0}$ at $\mathbf{p}_n$ is $\mathbf{h}(\mathbf{p}_n) + D\mathbf{h}(\mathbf{p}_n)\,(\mathbf{p} - \mathbf{p}_n) = \mathbf{0}$ where

$$D\mathbf{h}(\mathbf{p}) = D\left(\begin{bmatrix} h_1(p_1, \ldots, p_M) \\ \vdots \\ h_M(p_1, \ldots, p_M) \end{bmatrix}\right) = \begin{bmatrix} \frac{\partial h_1}{\partial p_1} & \cdots & \frac{\partial h_1}{\partial p_M} \\ \vdots & \ddots & \vdots \\ \frac{\partial h_M}{\partial p_1} & \cdots & \frac{\partial h_M}{\partial p_M} \end{bmatrix}$$

is the Jacobian matrix. Solving the linearized equation leads to the iteration

$$\mathbf{p}_{n+1} = \mathbf{p}_n - (D\mathbf{h}(\mathbf{p}_n))^{-1}\,\mathbf{h}(\mathbf{p}_n). \tag{2.9}$$

This is the multidimensional analog of (2.7).

2.5 Two-Parameter Exponential Fit

Let us illustrate (2.9) by estimating the parameters in the logistic curve (1.8). As before, $\mathbf{t}$ and $\mathbf{P}$ are the columns of the census data in Fig. 1.1 and $\mathbf{x} = \mathbf{t} - t_0$. In Sect. 1.7 we reformulated the logistic fit as a two-parameter exponential fit (1.9) which we performed using fminsearch. We will now minimize

$$g(a, c) = \|c + (1 - c)\,e^{-a\mathbf{x}} - \mathbf{y}\|^2, \quad \mathbf{y} = \frac{P_0}{\mathbf{P}} \tag{2.10}$$

using Newton's method; notice that the vector $\mathbf{y}$ in (2.10) is different from the one in Sect. 2.2.

Let $\mathbf{p}$ be the column vector with components a and c, in that order. Differentiating (2.10) (after expressing it in terms of the dot product) gives the gradient

$$\mathbf{h}(\mathbf{p}) = \begin{bmatrix} \frac{\partial g}{\partial a} \\ \frac{\partial g}{\partial c} \end{bmatrix} = \begin{bmatrix} 2\,(1 - c)\,(-\mathbf{x})\,e^{-a\mathbf{x}} \cdot \left(c + (1 - c)\,e^{-a\mathbf{x}} - \mathbf{y}\right) \\ 2\,\left(1 - e^{-a\mathbf{x}}\right) \cdot \left(c + (1 - c)\,e^{-a\mathbf{x}} - \mathbf{y}\right) \end{bmatrix}$$

and the Jacobian $J(\mathbf{p}) = D\mathbf{h}(\mathbf{p})$ with entries that simplify to

$$J_{1,1} = \frac{\partial^2 g}{\partial a^2} = 2\,(1 - c)\,\mathbf{x}^2\,e^{-a\mathbf{x}} \cdot \left(c + 2\,(1 - c)\,e^{-a\mathbf{x}} - \mathbf{y}\right)$$

$$J_{1,2} = J_{2,1} = \frac{\partial^2 g}{\partial c\,\partial a} = 2\,\mathbf{x}\,e^{-a\mathbf{x}} \cdot \left(2c - 1 + 2\,(1 - c)\,e^{-a\mathbf{x}} - \mathbf{y}\right)$$

$$J_{2,2} = \frac{\partial^2 g}{\partial c^2} = 2\,\left(1 - e^{-a\mathbf{x}}\right) \cdot \left(1 - e^{-a\mathbf{x}}\right)$$

Using the symmetry of the dot product, the $J_{1,1}$-entry could be expressed in terms of $\mathbf{x}\,e^{a\mathbf{x}}$ which would eliminate the need for computing $\mathbf{x}^2\,e^{-a\mathbf{x}}$. However, in the interest of keeping the code simple, we will not pursue that.

Notice that the Jacobian of **h** is the Hessian of g. This has several implications. Firstly, since mixed partial derivatives are equal, the Jacobian is *symmetric*: that is, it equals its transpose J^T—the matrix obtained from J by interchanging rows and columns. Secondly, at the minimum the Hessian should be *positive definite*. A reader familiar with linear algebra will know that symmetric matrices have real *eigenvalues* and orthogonal *eigenvectors*. Positive definite matrices have positive eigenvalues which can be checked using the `eig` command.

Newton's method (2.9) actually involves the inverse of the Jacobian. In principle, the inverse of a 2-by-2 matrix can be found using the general formula

$$\begin{bmatrix} a & b \\ c & d \end{bmatrix}^{-1} = \frac{1}{ad-bc} \begin{bmatrix} d & -b \\ -c & a \end{bmatrix}. \qquad (2.11)$$

However, it is best to avoid explicit computation of matrix inverses. In the following code we implement Newton's iteration as $\mathbf{p}_{n+1} = \mathbf{p}_n - \mathbf{q}_n$ where the "correction term" $\mathbf{q}_n$ is computed by solving $J(\mathbf{p}_n)\,\mathbf{q}_n = \mathbf{h}(\mathbf{p}_n)$. The starting values are given by (1.11) and (1.12); t and P must be set to columns of the data in Fig. 1.1.

```
x = t - t(1); y = P(1)./P;
x1 = x(1); x3 = x(end); x2 = .5*(x1 + x3);
y1 = y(1); y3 = y(end); y2 = interp1(x,y,x2);
T = .5*(x3 - x1);
a = log((y2 - y1)/(y3-y2))/T;
c = (y2 - exp(-a*x2))/(1 - exp(-a*x2));
p1 = [a;c]; g = zeros(2,1); J = zeros(2);
for n=1:20
    a = p1(1); c = p1(2);
    u = exp(-a*x); ux = u.*x; uxx = ux.*x;
    fy = c + (1-c)*u - y;
    g(1) = -(1-c)*dot(ux,fy);
    g(2) = dot(1-u,fy);
    J(1,1) = (1-c)*dot(uxx, c + 2*(1-c)*u - y);
    J(1,2) = dot(ux, 2*c - 1 + 2*(1-c)*u - y);
    J(2,1) = J(1,2);
    J(2,2) = dot(1-u,1-u);
    p2 = p1 - J\g;
    if norm(p2-p1) < eps('double')
        break
    else
        p1 = p2;
    end
end
```

2.5 Two-Parameter Exponential Fit

The loop terminates when $n = 7$ giving the following MLE:

$$\widehat{a} = 0.014\,299\,652\,238\,350 \text{ Year}^{-1}$$
$$\widehat{c} = 0.069\,010\,043\,489\,691$$
$$\widehat{\sigma} = 0.011\,889\,473\,445\,616$$

For comparison, fminsearch gives

$$a = 0.014\,300\,005\,370\,645 \text{ Year}^{-1}$$
$$c = 0.069\,026\,169\,226\,952$$
$$\sigma = 0.011\,889\,473\,551\,513$$

If we use $\widehat{a}$ and $\widehat{c}$ to compute the standard deviation for untransformed data, we get

$$\widetilde{\sigma} = \frac{1}{\sqrt{10}} P_0 \| \left(c + (1-c)e^{-a\mathbf{x}}\right)^{-1} - \mathbf{y}^{-1} \| = 0.033\,505\,928\,087\,395\, P_0.$$

The standard deviation for the logistic model is thus about 3.4% of the initial population size, which is only marginally better than for the natural growth model.

The eigenvalues of the Jacobian after the final iteration are

$$\lambda_1 = 0.067\,591\,383\,118\,391 > 0, \quad \lambda_2 = 3.787\,063\,464\,667\,862 \times 10^3 > 0.$$

Therefore we found a minimum, as confirmed by the contour plot in Fig. 2.6.

Again, the reader may wonder if using Newton's method is worthwhile when fminsearch seems to be just as good. We maintain that it is because, if nothing else, it validates the output of fminsearch with 15 digits of accuracy.

n	a_n	c_n
1	0.013920506359999	0.058843617857728
2	0.014160643931142	0.063547701185263
3	0.014297960505513	0.068997643891058
4	0.014299643419987	0.069009696691195
5	0.014299652238348	0.069010043489809
6	0.014299652238350	0.069010043489691

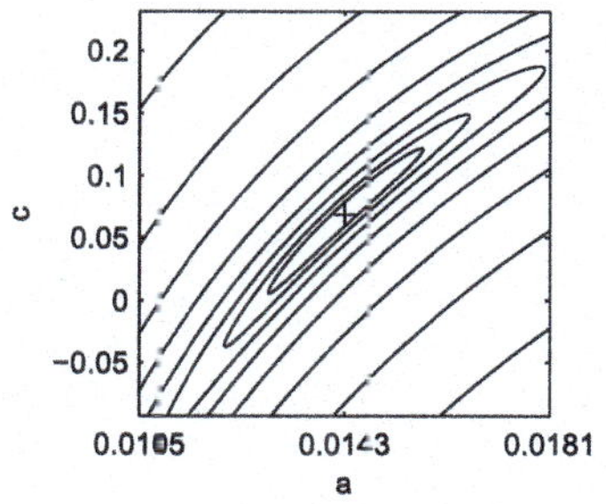

Fig. 2.6 Estimation of parameters a and c in (1.9) from the data in Fig. 1.1. The left panel shows Newton's iterations with starting values given by (1.11) and (1.12); the contour plot on the right confirms the minimum (marked with a cross)

2.6 Nonlinear Regression in 1-Norm

In Sect. 2.1 we showed that when the errors in the data are white noise, nonlinear least squares method gives maximum likelihood estimates and is therefore the best approach to parameter estimation. However, sometimes the structure of the noise is more complicated and calls for a different treatment.

If the data has outliers then, instead of minimizing the sum of squares (2.1), it may be better to minimize the sum of absolute values

$$g_1(p_1, \ldots, p_M) = \sum_{j=1}^{N} |f(x_j, p_1, \ldots, p_M) - y_j|. \tag{2.12}$$

Minimizer of (2.12) is called the 1-*norm best fit*.

Figure 2.7 compares NLS (the 2-norm best fit) with the 1-norm best fit for exponential data:

```
x = linspace(0,3,20);
y = 2*exp(-x) + .01*randn(size(x));
y(10) = 10*y(10);
```

To generate an outlier, we multiplied one of the components of **y** by 10.

Whereas the NLS fit is greatly perturbed by the outlier, the 1-norm fit barely notices it: optimization in 1-norm is more robust than in 2-norm. The catch is that we cannot

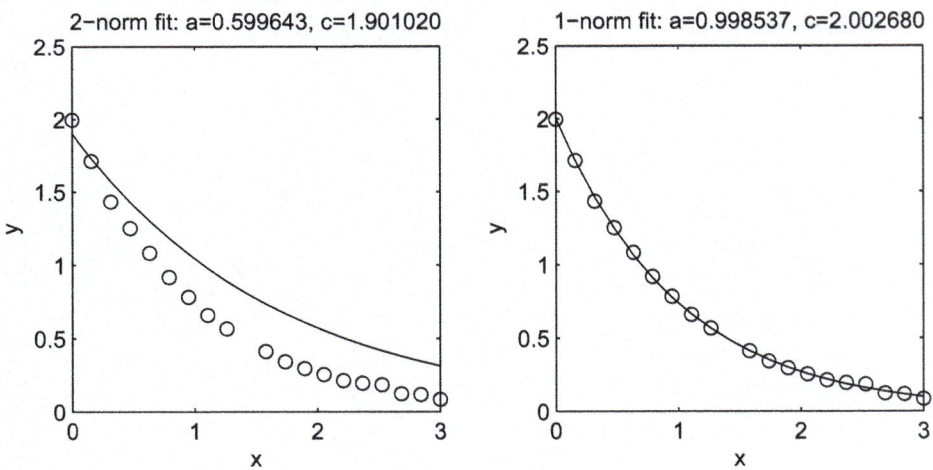

Fig. 2.7 Comparison of the 2-norm and the 1-norm exponential fits. The data was obtained by sampling $y = 2e^{-x}$ on $0 \le x \le 3$ and adding a small amount of white noise. To create an outlier, we scaled y_{10} by 10: this is why it appears to be missing

differentiate (2.12) and use Newton's method; to compute the 1-norm fit in Fig. 2.7 we used `fminsearch`.

2.7 Comments and Bibliography

Section 2.1 was inspired by [3] and the examples we used were adapted from that book, which we highly recommend.

Ancient Babylonians knew how to compute $\sqrt{2}$ to three sexagesimal places as evidenced by Tablet No. 7289 from the Yale Collection. It is very likely that they computed $\sqrt{2}$ by iterating

$$p_{n+1} = \frac{1}{2}\left(p_n + \frac{1}{p_n}\right).$$

For that reason, computation of $\sqrt{a}$ by iterating

$$p_{n+1} = \frac{1}{2}\left(p_n + \frac{a}{p_n}\right)$$

is called the *Babylonian method*; for more details, see [1].

As we mentioned in Sect. 2.2, Heron of Alexandria knew how to compute square roots of other numbers. Yet it was Newton who first understood the general method which now bears his name. In the Principia, Newton tested the method on the cubic equation $x^3 - 2x - 5 = 0$ before applying it to a non-polynomial equation from astronomy. Joseph Raphson, a contemporary of Newton, streamlined Newton's original procedure and because of that his name is often appended to Newton's, as in Newton-Raphson method.

A thorough and accessible explanation of multidimensional Newton's method can be found in [2]. That book is also an excellent general resource because it presents multivariate calculus as multidimensional and integrates it with linear algebra.

In Sect. 2.4 we mentioned eigenvalues and eigenvectors. These will be defined once we get to systems of linear ODE with constant coefficients in Chap. 7. Some prerequisite linear algebra notions will be clarified earlier, in Chap. 6.

Robustness of 1-norm to outliers is well known to statisticians and data scientists. However, minimization of the 1-norm of the residual does not have the same theoretical justification as NLS/MLE. Also, there are ways of dealing with outliers in the 2-norm and for that reason 1-norm fits are seldom used.

2.8 Exercises

1. As an homage to Newton, solve $x^3 - 2x - 5 = 0$ using Newton's method with $x_0 = 2$. Then map the basins of attraction of the roots of this cubic and compare them with Fig. 2.4.
2. Find a function $f(x)$ for which Newton's iteration

$$x_{n+1} = x_n - \frac{f(x_n)}{f'(x_n)}$$

with starting value $x_1 = 0$ produces the cycle

$$0, 1, 2, 3, 0, 1, 2, 3, 0, \ldots$$

Hint: set $f(x) = p(x) q(x) + 1$ where $p(x) = x(x-1)(x-2)(x-3)$ and q is a suitably chosen cubic polynomial. Does roundoff "liberate" Newton's method from this cycle? Generalize the problem to produce cycles of arbitrary length.

3. Sample a simple harmonic $f(t) = a \cos(\omega t) + b \sin(\omega t)$ on its period and add white noise to the sample values. The code for generating data may look as follows:

```
a = 3; b = -4; w = 3; T = 2*pi/w;
N = 500; t = linspace(0,T,N); s = .5;
y = a*cos(w*t) + b*sin(w*t) + s*randn(size(t));
```

Use NLS/MLE to estimate parameters a, b, ω, and the variance of the noise σ^2. Plot the histogram of the residual and confirm that it corresponds to $N(0, \sigma^2)$. Investigate the sensitivity of the parameter estimates to noise and the number of samples: does increasing the number of samples reduce the variance in the estimates? If so, how fast?

4. Generate the data as in the previous exercise and create one or more outliers. Compare the 2-norm estimates with 1-norm estimates in the manner of Fig. 2.7. Investigate the dependence of the 1-norm estimate on the number and strength of outliers: how many outliers of given strength does it take to ruin the 1-norm estimate?

5. This exercise requires strong programming skills, processing power, and perseverance. In Sect. 2.4 we estimated parameters of the logistic equations by finding the root of a certain nonlinear equation $\mathbf{h}(\mathbf{p}) = 0$. Map the basin of attraction for that root. Is it fractal?

References

1. Ezra Brown, Square roots from 1; 24, 51, 10 to Dan Shanks. Coll. Math. J. **30**(2), 82–95 (1999)
2. J.H. Hubbard, B.B. Hubbard, *Vector Calculus, Linear Algebra, and Differential Forms: A Unified Approach* (Prentice Hall, 1999)
3. Y. Pawitan, *In All Likelihood: Statistical Modelling and Inference Using Likelihood* (Oxford science publications, OUP Oxford, 2001)

Linearized ODE and Exponential Laws 3

In Chap. 1 we linearized ODE (1.1) to obtain the natural growth Eq. (1.3) solving which gave exponential law (1.5). As we saw in Sect. 1.5, despite not being physically realistic in the long term, exponential growth fits the U.S. census fairly well over the course of the 20th century.

Linearization is a common modeling gambit and many exponential laws originate as solutions of linearized ODE. In this chapter this is illustrated in four sections that are organized according to the same template. First, we set up a first order autonomous ODE

$$\frac{dx}{dt} = f(x) \qquad (3.1)$$

as a general model of some quantity x. Next we linearize (3.1), at a suitable point, to get a linear ODE

$$\frac{dx}{dt} = a\,x + b, \qquad (3.2)$$

whose solution, for a generic initial condition, is

$$x = \left(x_0 + \frac{b}{a}\right) e^{at} - \frac{b}{a}. \qquad (3.3)$$

Finally, we fit (3.3) to experimental data to confirm its validity.

The physical subject of Sect. 3.1 is convective heat transfer; in Sect. 3.3 it is an RC-circuit; in Sect. 3.5 it is one-dimensional motion with resistance; and in Sect. 3.7 it is atmospheric pressure which depends on altitude rather than time. The final Sect. 3.9 codifies the similarity between convective heat transfer and RC-circuits as electro-thermal analogy.

Rather than using the same notation in Sects. 3.1–3.7, we use the prevailing notation from the corresponding disciplines. As a result, each section has its own version of (3.1)–(3.3).

All exponential fits are performed using nonlinear least squares, as described in Chap. 2; to avoid repetition and save space, we present regression results without including code.

3.1 Convective Heat Transfer

Objects can exchange heat through conduction, convection, radiation, or some combination of all three mechanisms. In this section we construct a simple model of convective heat transfer to answer the following practical question:

> Suppose that we want to measure the temperature of a stream of liquid with a thermocouple. Once the thermocouple is inserted into the stream, it takes some time for its temperature T to equilibrate with the temperature of the liquid T_∞. How long should we wait before taking measurements?

A thermocouple is made by fusing the ends of two wires made from different alloys. It is that fused junction, which looks like a small metal droplet, that must be inserted into the stream, usually inside a protective sheath. Thinking of the thermocouple's junction as a small metal sphere with volume V, surface area A, density ρ, and specific heat capacity c, we can express the amount of thermal energy it stores as $c\,\rho\,V\,T$. The rate of change of that thermal energy is

$$c\,\rho\,V\,\frac{dT}{dt} = A\,\phi(T), \tag{3.4}$$

where $\phi(T)$ is heat flux—the rate of heat transfer per area. For large objects heat flux varies with location on the surface, but for a small metal sphere it only depends on the temperature of the sphere and the temperature of the environment. We did not list the latter as a variable in ϕ because we presume the temperature of the liquid T_∞ to be fixed, however it affects ϕ as a parameter.

Following the theme of the chapter, we will now linearize the right-hand side of (3.4) at $T = T_\infty$. Since there is no heat transfer without temperature difference, $\phi(T_\infty) = 0$. Therefore, for $T \approx T_\infty$, the heat flux is roughly proportional to the temperature gradient:

$$\phi(T) \approx \phi'(T_\infty)\,(T - T_\infty) = -h\,(T - T_\infty). \tag{3.5}$$

The symbol h is called the *convection heat transfer coefficient*; the minus sign in front of it signifies that heat flows from hot to cold.

Replacing $\phi(T)$ in (3.4) with its linear approximation (3.5), and solving for the derivative, leads to

$$\frac{dT}{dt} = -\tau^{-1}\,(T - T_\infty), \tag{3.6}$$

where the *time constant* τ is given by

3.2 The Cooling Light Bulb Experiment

$$\tau = \frac{c \rho V}{h A}. \tag{3.7}$$

In calculus and ODE textbooks Eq. (3.6) and its variants are referred to as *Newton's law of cooling*. In heat transfer, however, Newton's law of cooling is (3.5) and (3.6) is the energy balance involving thereof. Henceforth we will follow the heat transfer terminology.

The solution of (3.6) for a generic initial condition $T(0) = T_0$ is

$$T = T_\infty + (T_0 - T_\infty) e^{-\frac{t}{\tau}}. \tag{3.8}$$

Confusingly, this equation is also sometimes called Newton's law of cooling. Be this as it may, we can rewrite (3.8) as

$$\frac{T_0 - T(t)}{T_0 - T_\infty} = 1 - e^{-\frac{t}{\tau}},$$

and designate the fraction on the left as a measure of completeness of heat transfer: the closer it is to unity, the closer the point of equilibration. By that measure, when $t = 5\tau$ the heat transfer process is more than 99.3% complete. Therefore the standard rule for measuring temperature with thermocouples (and other devices) is to wait 5 time constants before taking the first measurement.

It may appear that the time constant τ can be computed directly from Eq. (3.7), but, while the material and geometric properties c, ρ, V and A can be measured or looked up in tables, the same cannot be said about the convection heat transfer coefficient h. In practice, τ is determined from experimental data and h is found using (3.7).

One of the main purposes for finding h is to compute the *Biot number*

$$\text{Bi} = \frac{h V}{k A}. \tag{3.9}$$

This non-dimensional ratio compares convection to internal conduction whose heat transfer coefficient k is in the denominator. If $\text{Bi} \ll 1$ the temperature of the object may be presumed uniform; otherwise, conduction should be taken into account by replacing ODE (3.6) with a PDE.

3.2 The Cooling Light Bulb Experiment

Equation (3.8) predicts that the temperature of a small object immersed in a steady stream of liquid exponentially approaches the temperature of the liquid. To test that, we measured the temperature of a 40 W incandescent light bulb cooled by a steady draft of air.

The bulb was turned off when its temperature was about 275 °F, at which point a desktop fan was turned on to generate draft; the fan was positioned to ensure strong, but not turbulent airflow. The temperature of the bulb was taken at the point on the globe (glass part) that is farthest from the base (metal part). We used PASCO™ PS-2125 temperature sensor

t (s)	T_∞ (°F)	T (°F)
0	80.67	274.84
10	80.51	223.34
20	80.51	184.33
30	80.60	156.09
40	80.67	135.18
50	80.67	120.02
60	80.65	109.06
70	80.64	101.08
80	80.65	95.34
90	80.67	91.29
100	80.69	88.41
110	80.73	86.32
120	80.74	84.87
130	80.74	83.77
140	80.76	83.01

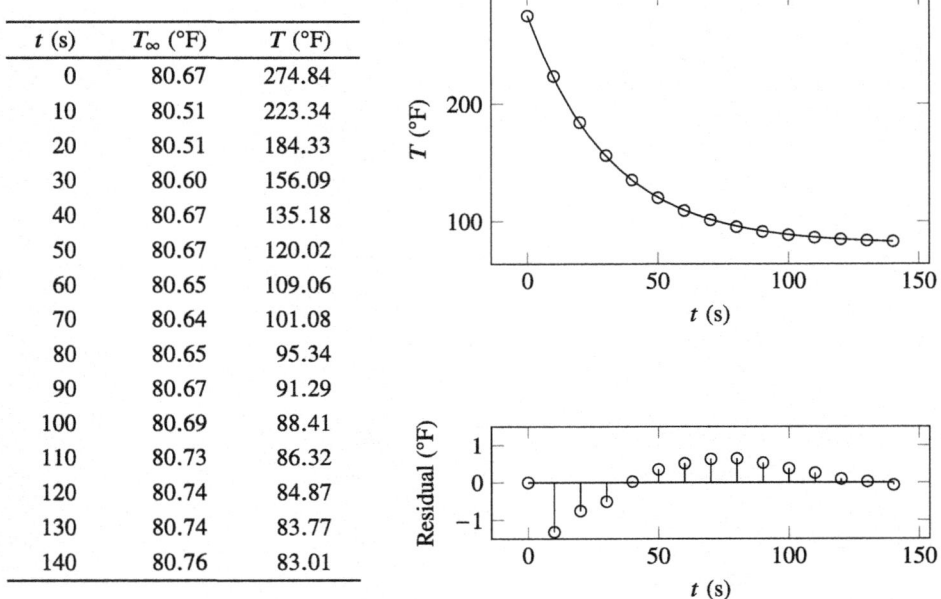

Fig. 3.1 Temperature of a cooling light bulb

which has accuracy ± 0.9 °F and resolution (smallest change that the sensor is capable of detecting) 0.018 °F; the temperature of air was measured in close proximity to the bulb with the same type of sensor.

The temperatures of air and the bulb were recorded at the sampling rate of 10 Hz until their difference fell below experimental accuracy set to 1.8 °F. Figure 3.1 shows 15 of the measurements and the exponential fit that was computed using the entire data set of 1474 observations.

To compute the exponential fit, we estimated T_∞ as the mean of air temperature and set the first measurement of bulb's temperature as T_0. To estimate the time constant, we minimized

$$g(\tau) = \|T_\infty + (T_\infty - T_0)\,e^{-\frac{t}{\tau}} - \mathbf{T}\|^2$$

following the methodology of Sect. 2.3—we found the root of $g'(\tau)$ using Newton's method with the starting value $\tau_1 = 31.338987010468319$ s obtained from linear regression. It took Newton's method only 4 iterations to converge to $\hat{\tau} = 31.501746286493649$ s, as befits a quadratically convergent method.

For the entire data set the maximum absolute error is slightly less than 1.5 °F which is below the experimental accuracy of 1.8 °F; the mean of the absolute error is about 0.45 °F, which is almost exactly a half of the 0.9 °F accuracy of the sensor.

We computed $\hat{\sigma} \approx 0.55\,°\text{F}$, but, since the stem plot of the residual in the bottom right corner of Fig. 3.1 does not resemble white noise, $\hat{\sigma}$ should not be interpreted as the standard deviation of a normal distribution. The value of $\hat{\sigma}$ does indicate that the discrepancy between the data and the model is acceptable, given the accuracy of the equipment. Yet the plot of the residual makes it clear that the simple exponential law (3.8) does not fully describe the data.

One likely reason why (3.8) deviates from the data is the presence of other forms of heat transfer. To see if internal conduction was the culprit, we computed the convection heat transfer coefficient and the Biot number. Idealizing the light bulb as a thin spherical shell made from soft soda-lime glass, we set

$$\frac{V}{A} = 5 \times 10^{-4}\,\text{m}, \quad \rho = 2500\,\text{kg}\,\text{m}^{-3}, \quad c = 870\,\text{J}\,\text{kg}^{-1°}\text{K}^{-1},$$

and used Eq. (3.7) to find

$$h = \frac{c\rho V}{\tau A} \approx 35\,\text{W}\,\text{m}^{-2°}\text{K}^{-1},$$

which is in the range $10\text{–}500\,\text{W}\,\text{m}^{-2°}\text{K}^{-1}$ typical for forced air convection. We then used (3.9) with $k = 1.06\,\text{W}\,\text{m}^{-1°}\text{K}^{-1}$ to find $\text{Bi} \approx 0.0162$.

Since Biot number is small, thermal conduction within the bulb can be ignored. With this said, there are thermal losses due to conduction through the base of the bulb and the metal sheath of the temperature sensor. There is also thermal radiation, irregularity of air flow, and other factors that (3.6) does not take into account. Nevertheless, we adjudicate the exponential law (3.8) to be an adequate description of the temperature of the light bulb and therefore the linearized model (3.6) is valid in this case.

3.3 RC-Circuit

The RC-circuit of this section is electric, but, in order to gain intuition, we will first discuss its hydraulic analog shown in Fig. 3.2.

A hydraulic resistor is a sharp bend in a pipe, a constriction, or any obstacle that impedes flow. Actually, all pipes, even straight and wide ones, have some resistance due to viscous friction. However, for the purpose of analysis, pipes are idealized: a real pipe is an ideal pipe plus a hydraulic resistor.

A hydraulic capacitor can be made by dividing a sealed vessel into two chambers with a flexible membrane. The shape of the vessel is immaterial—in Fig. 3.2 we show a cross-section of a cylinder divided in the middle, but it could be a box, or a sphere. Just like pipes, capacitors are assumed to have no internal resistance: all resistance in an RC-circuit is "lumped" into the resistor.

The dots in Fig. 3.2 symbolize air which the pump P can move from one chamber of the capacitor to another. On the left side of the figure the pump is off, there is equal pressure in

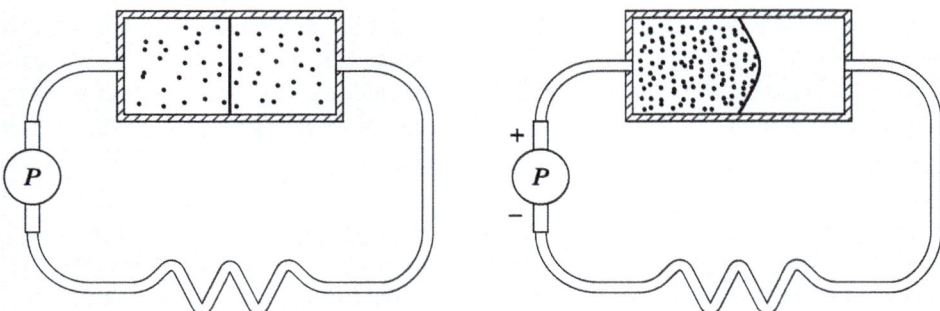

Fig. 3.2 Hydraulic RC-circuit

both chambers of the capacitor and the membrane is relaxed—the capacitor is "discharged". On the right side the pump creates pressure gradient indicated with the + and − signs, the chambers are at different pressures and the membrane is stretched—the capacitor is "charged".

If the pump is shut off once the capacitor is charged, the membrane will expel air from the chamber with high pressure into the chamber with low pressure. Owing to the resistor, this will happen gradually: the higher the resistance the slower the discharge.

Now let us consider an electric RC-circuit. On the left side of Fig. 3.3 the circuit consists of a resistor R, capacitor C, and a switch S which is shown in the open position; on the right we added a battery E and closed the switch.

An electric resistor, just like a bent pipe, impedes flow, only this time it is the flow of charge—that is why it is drawn as a zigzag. All other circuit elements are idealized to have no resistance, so the resistor R accounts for all of the resistance in the circuit, including that of the wiring.

An electric capacitor is, in essence, two metal plates separated by a layer of dielectric—hence the symbol that looks like a break in the circuit. The capacitor C is non-polarized, meaning that its plates are identical.

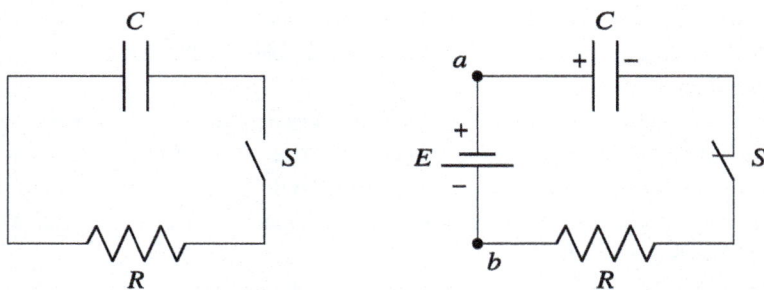

Fig. 3.3 Electric RC-circuit

3.3 RC-Circuit

The battery E is a pump for electrons: it creates voltage drop—analog of pressure drop—that forces electrons to migrate from one plate of the capacitor to another. As a result, the plates acquire equal but opposite charges thereby creating an electric field in the gap between them. That field is like a stretched membrane in a hydraulic capacitor: if we open the switch on the right of Fig. 3.3, replace the battery with a wire (short), as shown on the left, and close the switch, the electric field will push the electrons from the negatively charged plate to the positively charged plate until both plates are electrically neutral.

In order to model the electric RC-circuit, we need to define its components by their current-voltage characteristics. As is customary in circuit analysis, we will use I for current, V for voltage, and Q for charge on the positive plate of the capacitor.

Mathematically, a resistor is an element whose current-voltage characteristic is purely algebraic: $I = f(V)$. Thus, if we know the voltage drop across the resistor, we know the current that flows through it. We will assume the converse as well, so that we can write $V = f^{-1}(I)$ when required.

For a capacitor, the current is proportional to the rate of change of voltage:

$$I = C \frac{dV}{dt}. \tag{3.10}$$

In physics Eq. (3.10) is usually derived by differentiating *Volta's law*: $Q = CV$; current is, by definition, the rate of change of charge.

Finally, we are going to assume that the battery E is an ideal voltage source. This means that the voltage across it is constant regardless of the current: $V = E$.

Let V be the voltage drop across the capacitor and let I be the current through the resistor. If we traverse the circuit on the right side of Fig. 3.3 from a to b clockwise, the voltage will drop first across the capacitor by V, and then across the resistor by $f^{-1}(I)$. Yet, if we move counterclockwise, the voltage drop will equal that created by the battery. According to *Kirchhoff's voltage law* (KVL), the sum of voltage drops along any branch connecting two nodes is the same regardless of the branch, therefore:

$$V + f^{-1}(I) = E. \tag{3.11}$$

Since the circuit in Fig. 3.3 is a single loop, by *Kirchhoff's current law* (KCL) the current through the resistor must equal the current through the capacitor (and the battery). Accordingly, we can replace I in (3.11) with the expression (3.10). After solving for the derivative, the resulting ODE is

$$C \frac{dV}{dt} = f(E - V). \tag{3.12}$$

We now stipulate that the resistor R has linear current-voltage characteristic

$$I = f(V) = \frac{V}{R}. \tag{3.13}$$

With the assumption of *Ohm's law* (3.13), Eq. (3.12) can be written as

$$\frac{dV}{dt} = \frac{1}{RC}(E - V). \tag{3.14}$$

The solution of (3.14) with generic initial condition is

$$V = E + (V_0 - E)\, e^{-\frac{t}{RC}}. \tag{3.15}$$

For the RC-circuit on the left side of Fig. (3.3) (with the switch closed), we simply need to set $E = 0$.

3.4 The Discharging Capacitor Experiment

To test Eq. (3.15), we discharged a $2.2\,\mu\text{F}$ capacitor through a $10\,\text{k}\Omega$ resistor and measured the voltage across the capacitor with Digilent Analog Discovery™ oscilloscope. The measurements were collected for 100 ms at the rate of 50 kHz. Figure 3.4 shows 20 observations and the exponential fit computed using the entire data set of 5000 observations.

t (ms)	V (V)
0	13.0330
5	10.4508
10	8.3795
15	6.7350
20	5.4159
25	4.3452
30	3.4775
35	2.7987
40	2.2668
45	1.7945
50	1.4551
55	1.1822
60	0.9372
65	0.7623
70	0.6049
75	0.4929
80	0.4159
85	0.3179
90	0.2480
95	0.2165

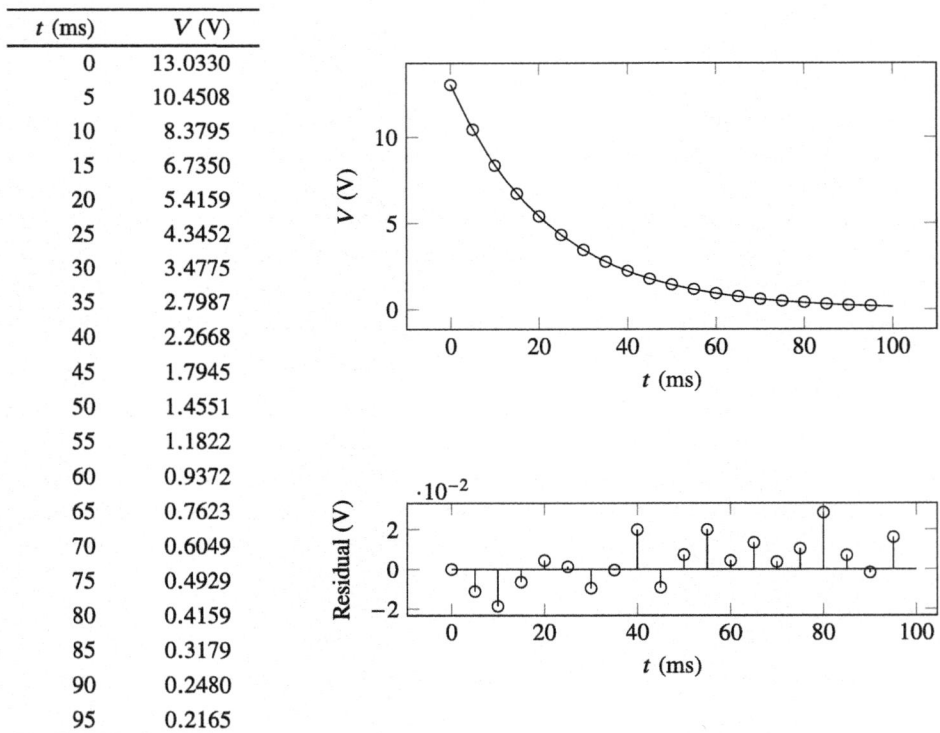

Fig. 3.4 Voltage across a capacitor discharging through a resistor

3.4 The Discharging Capacitor Experiment

According to (3.15), in the absence of the battery the capacitor voltage is

$$V = V_0 \, e^{-\frac{t}{RC}}. \tag{3.16}$$

Since the initial voltage is part of the data, the only parameter that needs to be estimated is the time constant $\tau = RC$.

For a $10\,\mathrm{k}\Omega$ resistor and a $2.2\,\mu\mathrm{F}$ capacitor the time constant should be $22\,\mathrm{ms}$. However, since the tolerance on the resistor is 5% and the tolerance on the capacitor is 10%, the true time constant can be anywhere in the range of 18.81–$25.41\,\mathrm{ms}$. Using NLS, we found $\widehat{\tau} = 22.753927046605931\,\mathrm{ms}$ which is close to the nominal value.

For the entire data set, the maximum error of (3.16) is $52\,\mathrm{mV}$ while the mean error is $11\,\mathrm{mV}$. We found $\widehat{\sigma} \approx 14\,\mathrm{mV}$ and compared the histogram of the residual with the normal distribution $N(0, \widehat{\sigma})$ in Fig. 3.5.

The histogram is close to a normal distribution but with mean of about $3.4\,\mathrm{mV}$ and standard deviation that is slightly lower than $\widehat{\sigma}$. The plot of the residual shows that it is not pure white noise.

Based on the results shown in Figs. 3.4 and 3.5, we deem the exponential law (3.16) and the linear model (3.14) satisfactory. Still, the model can be improved by taking into account *probe loading* and *soakage*.

Attaching a probe of an oscilloscope to a circuit changes the circuit: this is probe loading. A simple model of the probe is a $1\,\mathrm{M}\Omega$ resistor connected in parallel, but a more precise model would also include the probe's capacitance which should be on the order of a few pF—we measure voltage in an RC-circuit with another RC-circuit.

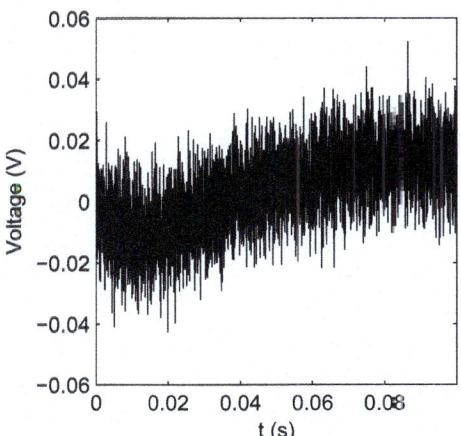 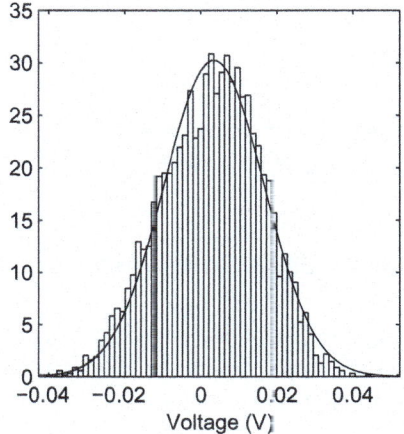

Fig. 3.5 Plot of the residual for RC-circuit data (left) and its histogram compared with $N(0, \widehat{\sigma})$ (right)

Real capacitors do not fully discharge due to dielectric absorbtion, or soakage: they have "memory" which acts as a weak voltage source. To take that into account, we should regard E in (3.15) as a parameter instead of setting it to zero.

We should also mention that the probe of an oscilloscope, being an antenna, is affected by the omnipresent 60 Hz "hum" from electric mains. Even if a probe is not attached to anything, the oscilloscope will trace a 60 Hz sine wave with amplitude of about 2 mV.

3.5 One-Dimensional Motion with Resistance

Consider an object moving along a straight line under the influence of a force that depends only on the object's velocity v. According to Newton's second law,

$$\frac{dv}{dt} = f(v), \tag{3.17}$$

where f is the force divided by the object's mass. We will write the linearization of (3.17) at $v = 0$ as

$$\frac{dv}{dt} = p - r\,v, \tag{3.18}$$

with the usual understanding that $p = f(0)$ and $r = -f'(0)$ are positive parameters; the term $-r\,v$ thus corresponds to friction or *drag*.

The solution of (3.18) is

$$v = \frac{p}{r} + \left(v_0 - \frac{p}{r}\right) e^{-rt}. \tag{3.19}$$

Integrating (3.19) gives the distance traveled by the object

$$s = \frac{p}{r} t + \left(v_0 - \frac{p}{r}\right) \frac{1 - e^{-rt}}{r}. \tag{3.20}$$

As $t \to \infty$ the velocity approaches *terminal velocity* $v_\infty = \frac{p}{r}$ and the distance approaches its linear asymptote with slope v_∞.

3.6 Usain Bolt's World Record

Equation (3.18) may be used to model the velocity of a sprinter. The parameter p is then specific thrust (propulsive force-per-mass) generated by the athlete while r accounts for dissipation of energy within muscles and ligaments—air resistance is negligible in comparison.

Since sprinters start from rest, the initial velocity in Eqs. (3.19) and (3.20) should be set to zero. The distance function is therefore

$$s = \frac{p}{r}\left(t - \frac{1 - e^{-rt}}{r}\right). \tag{3.21}$$

3.6 Usain Bolt's World Record

s (m)	t (s)
0	0.00
10	1.88
20	2.88
30	3.78
40	4.64
50	5.47
60	6.29
70	7.10
80	7.92
90	8.74
100	9.58

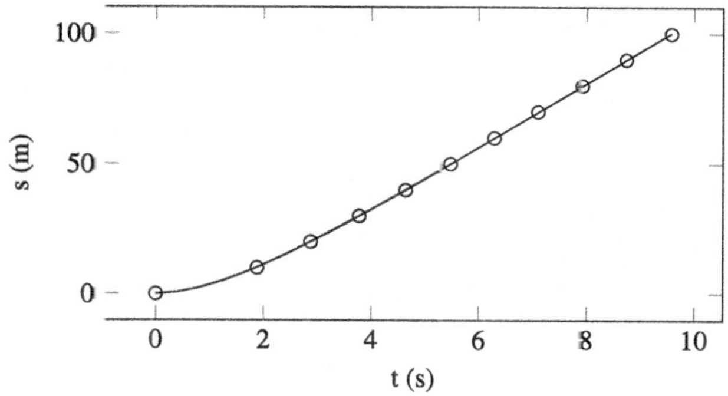

Fig. 3.6 Usain Bolt's 100 m world record (IAAF)

At athletic events, distances in 100 m dashes are officially timed in 10 m increments.

On August 16, 2009 then 22 year old Usain Bolt from Jamaica set the world record in men's 100 m dash at the International Association of Athletics Federations (IAAF) World Championships in Berlin. Bolt's time of 9.58 s was a dramatic improvement of the 9.74 s record set by his countryman, Asafa Powell, on September 9, 2007 at the IAAF Grand Prix in Rieti, Italy; so far it has not been challenged.

Figure 3.6 shows the official IAAF data for Usain Bolt's 100 m world record.

We fitted (3.21) to this data by refining the output of fminsearch using Newton's method. With the resulting estimates

$$\widehat{p} = 8.492425638441933 \text{ m/s}^2, \quad \widehat{r} = 0.688382862694927 \text{ s}^{-1}$$

the maximum error of (3.21) is less than 29 cm and the standard deviation $\widehat{\sigma}$ is about 14 cm.

Figure 3.7, produced from IAAF data aggregated by www.athletefirst.org, shows the distribution of parameters p and r from 200 races for men and for women. Notice that Usain Bolt's world record parameters are not outliers, as one would suspect.

The range of specific thrust p is roughly the same for both sexes, however, the internal resistance r tends to be higher for women. The current female 100 m world record of 10.49 s was set by Florence Griffith-Joyner at United States Olympic Trials in Indianapolis, Indiana, on July 16, 1988.

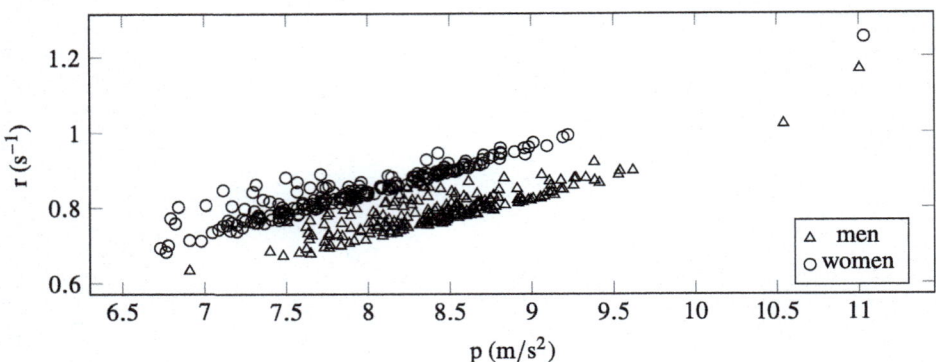

Fig. 3.7 Distribution of parameters p and r for men and women

3.7 Atmospheric Pressure

This section will serve as a counterpoint to the preceding three by showing that exponential laws derived from linearized ODE are not always unequivocal. Here we will also use altitude h (height above sea level) as an independent variable, rather than time t.

Let $P(h)$, $\rho(h)$, and $g(h)$ be atmospheric pressure, air density, and acceleration due to gravity, respectively. Consider a static column of air with cross-section A, and imagine it being composed of horizontal layers with thickness Δh. Consider now the forces acting on the layer at altitude h: the upward pressure force $P(h)\,A$ due to the layers below, the downward pressure force $P(h+\Delta h)\,A$ due to the layers above, and the weight of the layer itself $g(h)\,\rho(h)\,\Delta h\,A$. Equilibrium requires the net force to be zero

$$P(h)\,A - P(h+\Delta h)\,A - g(h)\,\rho(h)\,\Delta h\,A = 0,$$

whence follows

$$\frac{P(h+\Delta h) - P(h)}{\Delta h} = -g(h)\,\rho(h).$$

Taking the limit, as $\Delta h \to 0$, leads to the equation of hydrostatic equilibrium

$$\frac{dP}{dh} = -g(h)\,\rho(h), \tag{3.22}$$

which we will linearize by assuming the following: (i) the acceleration due to gravity is constant: $g(h) = g(0) = g_0$; (ii) air obeys the ideal gas law which may be stated as

$$P = \frac{RT}{M}\rho, \tag{3.23}$$

where T is temperature, M is molar mass of air, and R is the universal gas constant; (iii) the atmosphere is isothermal: $T = T(0) = T_0$. With these assumptions (3.22) becomes

$$\frac{dP}{dh} = -\frac{P}{\eta}, \quad \eta = \frac{R T_0}{M g_0}. \tag{3.24}$$

The *characteristic altitude* η in (3.24) is the analog of the time constant.
Setting

$$T_0 = 293.15 \text{ K}, \quad R = 8.31 \text{ Jmol}^{-1}\text{K}^{-1}, \quad M = 28.96 \text{ g/mol}, \quad g_0 = 9.81 \text{ m/s}^2$$

gives $\eta \approx 8575$ m meaning that for 20 °C atmosphere the pressure should roughly decrease by half for each 6 km of elevation. Of course, since temperature drops with altitude, η is likely to be lower.

Integrating (3.24) gives the *barometric law*

$$P = P_0 \, e^{-\frac{h}{\eta}}. \tag{3.25}$$

The atmospheric pressure at sea level is $P_0 = 101$ kPa.

3.8 NASA 1976 Standard Atmosphere Model

To validate (3.25) we matched it against the data from NASA 1976 standard atmosphere model. In Fig. 3.8 the altitude is geopotential but it is almost the same as the altitude above sea level.

We set $P_0 = 1013.25$ mbar and estimated $\widehat{\eta} \approx 7591$ m which corresponds to constant atmospheric temperature $T_0 = 260$ K or -13 °C. As the plot of the residual shows, for elevations under 15 km exponential law (3.25) gives errors on the order of 20 mbar. For altitudes below 10 km this corresponds to less than 3% relative error, however for higher altitudes the relative error quickly grows, exceeding 16% at 15 km.

The most problematic assumption behind (3.25) is that the atmosphere is isothermal—it is not: the temperature in the troposphere—the layer of the atmosphere below 11 km—decreases by about 6.5 °C with each kilometer of altitude. The temperature of the next layer, the tropopause, is nearly constant, however, and there the exponential model is more appropriate.

Despite its flaws, the barometric law (3.25) is useful at small altitudes when errors of a few percent are acceptable. In the past, it was the main working principle of the altimeter—a device that measures static pressure outside an aircraft and converts it into altitude.

h (km)	P (mbar)
0	1,013.25
1	898.74
2	794.95
3	701.08
4	616.40
5	540.19
6	471.81
7	410.60
8	355.99
9	307.42
10	264.36
11	226.32
12	193.30
13	165.10
14	141.01
15	120.44

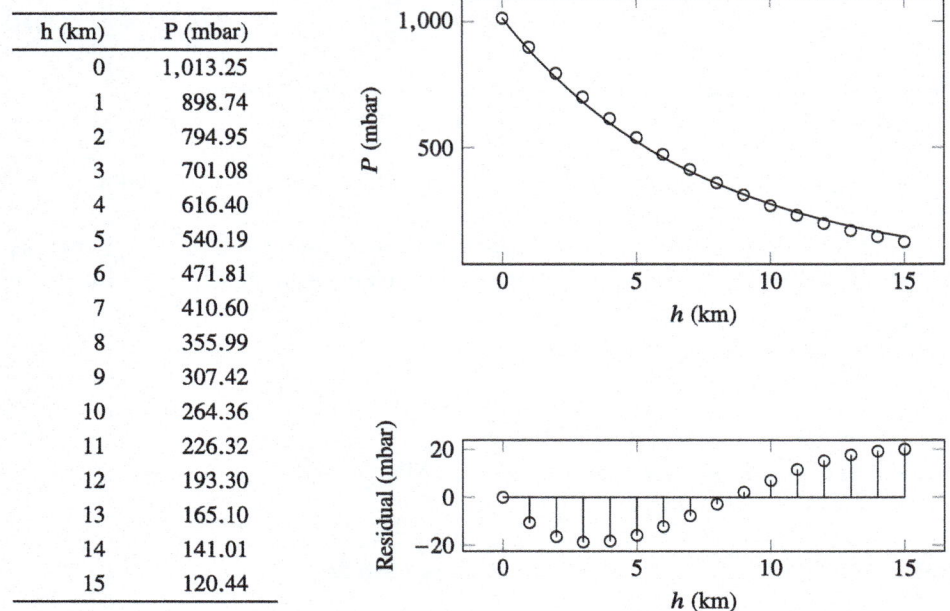

Fig. 3.8 NASA 1976 standard atmosphere model

3.9 Electro-Thermal Analogy

We opened Sect. 3.3 with a discussion of a hydraulic analog of an electric RC-circuit. That analogy is purely qualitative: mathematically, the hydraulic RC-circuit depicted in Fig. 3.2 is different from its electric counterpart in Fig. 3.3 because flow rate-pressure characteristics of hydraulic elements are more complicated than the corresponding current-voltage characteristics of electric elements. In this section we want to draw the attention of the reader to a much closer analogy between electric RC-circuits and thermal RC-circuits.

Compare Eqs. (3.15) and (3.6): the difference is in notation only. Changing the latter using the correspondence

$$V \leftrightarrow T, \quad E \leftrightarrow T_\infty, \quad RC \leftrightarrow \tau \tag{3.26}$$

transforms (3.15) into (3.6) and vice versa. Thus, from the mathematical point of view, there is no difference between a light bulb cooling in a draft and a capacitor discharging through a resistor. In fact, a cooling light bulb is a thermal capacitor discharging through a thermal resistor.

Comparison of the electric time constant RC with its thermal counterpart given by (3.7) leads to the identification

$$R \leftrightarrow \frac{1}{hA}, \quad C \leftrightarrow cm, \tag{3.27}$$

where we wrote mass m in place of the product of density with volume to avoid notational conflict between volume and voltage. Equations (3.26)–(3.27) form what we will call an electro-thermal analogy.

Thermal RC-circuits are drawn using the same symbols as in electric RC-circuits. The cooling light bulb from Sect. 3.1 can thus be diagrammed as the circuit in the right panel of Fig. 3.3, but with thermal resistance R and thermal capacitance C given by (3.27) and the battery's temperature set to $E = T_\infty$.

The analogy between electric RC-circuits and convective heat transfer can be extended to conductive heat transfer by dividing solids into small "elements" for which the Fourier's law of heat conduction may be approximated with a relation similar to Newton's law of cooling. Radiative heat transfer can also be modeled with electric RC-circuits but that requires nonlinear circuit elements.

3.10 Comments and Bibliography

Heat transfer has a special place in science because of its profound influence on other disciplines. To give the reader a sense of that influence, we will briefly describe some of the milestones, beginning with Newton's law of cooling.

In May of 1701 Newton presented to the Royal Society a number of temperature measurements, including melting points of several metals and alloys. If that does not sound remarkable, imagine what it was like to measure temperature in the late 17th century: the thermometers of the time were glass tubes filled with spirits of wine, wholly unsuitable for dipping into molten lead and antimony!

Newton replaced alcohol with linseed oil which has both a higher boiling point and a higher coefficient of expansion; for some reason, he did not choose mercury even though he had it on hand—the invention of mercury-in-glass thermometer (by Fahrenheit) would have to wait until 1714. While the boiling point of linseed oil is around 600 °F, it smokes at 225 °F and begins to change color and viscosity at 400 °F. To get around that, Newton supplemented his linseed oil thermometer with an iron block thermometer which he described to the Royal Society as follows:

> ...there was heated a pretty thick piece of iron red-hot, which was taken out of the fire with a pair of pincers, which were also red-hot, and laid in a cold place, where the wind blew continually upon it, and putting on it particles of several metals, and other fusible bodies, the time of its cooling was marked, till all the particles were hardened, and the heat of the iron was equal to the heat of the human body; then supposing that the excess of the degrees of the heat of the iron, and the particles above the heat of the atmosphere, found by the thermometer, were in geometrical progression, when the times are in an arithmetical progression, the several degrees of heat were discovered; the iron was laid not in a calm air, but in a wind that blew uniformly upon it, that the air heated by the iron might be always carried off by the wind, and

the cold air succeed it alternately; for thus equal parts of air were heated in equal times, and received a degree of heat proportional to the heat of the iron;

It is clear from this description that Newton inferred melting points from cooling times. In recognition of that, the main law of convective heat transfer is named in his honor. By the way, Newton and his contemporaries did not distinguish between temperature and heat. That distinction was first made by the Scottish chemist Joseph Black in 1760 but became widely known only in the early 1800's.

The concept of heat transfer rate q and the convection heat transfer coefficient h were introduced in 1822 by Fourier in his monumental *Théorie Analytique de la Chaleur* (*The Analytical Theory of Heat*), and, because of that, the law of cooling is sometimes erroneously attributed to him. Fourier is rightfully credited with the theory of heat conduction which is epitomized by *Fourier's law*

$$\mathbf{q} = -k \, \nabla T. \tag{3.28}$$

However, as far as modeling forced convection is concerned, that honor is unquestionably Newton's. For more on the early history of heat transfer and the contributions of Newton, Fourier, Biot, and others consult [14, 15, 17], and the references therein.

Newton was undoubtedly aware of radiative heat transfer but did not know how to quantify it—the Stefan-Boltzmann law would be discovered experimentally by Josef Stefan only in 1879 and deduced theoretically by Ludwig Boltzmann in 1884. With radiation unaccounted for, Newton's measurements above 450 °F (the melting point of tin) are lower than their true values. In [8] the temperature of Newton's "red-hot iron" is modeled with

$$c \, \rho \, V \, \frac{dT}{dt} = -h \, A \, (T - T_\infty) - \epsilon \, \sigma \, A \, \left(T^4 - T_\infty^4\right), \tag{3.29}$$

where the second term on the right-hand side is the rate of radiative heat transfer given by the Stefan-Boltzmann law: $\sigma = 5.67 \times 10^{-8}\,\mathrm{W\,m^{-2\circ}K^{-4}}$ is the Stefan-Boltzmann constant; ϵ is the non-dimensional emissivity of the surface whose value is between 0 and 1. As the authors of [8] show, when thermal radiation is taken into account Newton's iron block thermometer becomes much more accurate.

Fourier was greatly inspired by Newton and, in turn, inspired others, like Ohm, Kirchhoff, and Fick, to seek laws analogous to Fourier's law (3.28).

Georg Ohm discovered (3.13) in 1827, five years after reading Fourier's *Théorie Analytique de la Chaleur*. Around 1845 Kirchhoff reformulated Ohm's law in vector form

$$\mathbf{J} = \sigma \, \mathbf{E}, \tag{3.30}$$

where $\mathbf{J}$ is current density, $\mathbf{E}$ is electric field, and σ is *conductivity*—the inverse of resistance. When electric field is proportional to the gradient of voltage, as it often is, Ohm's law (3.30) is mathematically identical to Fourier's law (3.28). Reflecting on that, Ohm wrote in [11]:

3.10 Comments and Bibliography

The form and treatment of the differential equations [which describe electric current] thus obtained are so similar to those given for the propagation of heat by Fourier and Poisson, that even if there existed no other reasons, we might with perfect justice draw the conclusion that there exists an intimate connection between these natural phenomena; and this similarity increases as we continue to pursue the subject.

In 1855 Adolph Fick formulated his law of diffusion

$$\mathbf{q} = -D\,\nabla c, \tag{3.31}$$

where $\mathbf{q}$ is mass flux, c is concentration, and D is diffusivity. Echoing Ohm, Fick wrote in [2]:

It was quite natural to suppose, that this law for the diffusion of a salt in its solvent must be identical with that, according to which the diffusion of heat in a conducting body takes place; upon this law Fourier founded his celebrated theory of heat, and it is the same which Ohm applied with such extraordinary success, to the diffusion of electricity in a conductor.

We should add that, unlike Fourier's law and Fick's law, Ohm's law was not an instant success. In fact, it was met with such scepticism that Ohm had to resign from his university position in Cologne! In 1827 Ohm did not have a good understanding of electromagnetism and relied instead on what we can call the original electro-thermal analogy. The vagueness of Ohm's reasoning invited doubt including the following withering criticism from Maxwell in [10]:

Ohm, misled by the analogy between electricity and heat, entertained an opinion that a body when raised to a high potential becomes electrified throughout its substance, as if electricity were compressed into it, and was thus by means of an erroneous opinion led to employ the equations of Fourier to express the true laws of conduction of electricity through a long wire, long before the real reason of the appropriateness of these equations had been suspected.

Maxwell may have a point. However, the fact of the matter is that today, together with Fourier's and Fick's laws, Ohm's law forms the triumvirate of gradient laws of mathematical physics.

One of the first electric RC-circuits—the subject of Sect. 3.3—was inadvertently created in 1745 by Ewald Georg von Kleist when he used a metal wire to guide static electricity from an electrified glass ball into a glass jar filled with alcohol. Kleist discovered that touching the wire with one hand while holding the jar with the other produced a big spark—unwittingly, he shorted a capacitor formed by the palm of his hand and the jar's interior surface.

In November of 1746 the news of Kleist's big sparks reached the University of Leyden professor Pieter van Musschenbroek who repeated Kleist's experiments with his student Andreas Cunaeus, but with water in lieu of alcohol. Both van Musschenbroek and Cunaeus received electric shocks of such severity that the former wrote to René-Antoine Ferchault de Réaumur (the inventor of Réaumur temperature scale):

> I would like to tell you about a new but terrible experiment, which I advise you never to try yourself, nor would I, who have experienced it and survived by the grace of God, do it again for all the kingdom of France.

Not only was van Musschenbroek's dire warning summarily ignored, his apparatus was copied and sold far and wide, as the *Leyden jar*. By mid-18th century Leyden jars became a staple in European scientific laboratories and were gradually spreading throughout the world. Some of them crossed the Atlantic where they attracted the attention of Benjamin Franklin.

Franklin had a collection of Leyden jars, a few of which were lined with foil. In the course of his famous researches in electricity Franklin discovered, among other things, that the charge was stored not in the liquid but on the surface of the glass. Realizing that he did not need the liquid, Franklin replaced the jars in some of his experiments with flat pieces of glass coated with foil on either side—*Franklin squares*. Those were the ancestors of capacitors which today are found in nearly all electronic devices.

Franklin did not know that the carriers of charge in metals were negatively charged electrons—that was established by Thomson in 1897. Assuming that the mobile charges were positive, Franklin reasoned that they moved from positive to negative, whereas in reality electrons move from negative to positive. Rather than correct Franklin, physicists call his current *conventional* as opposed to real or physical. In Franklin's defense, it usually does not matter whether real or conventional current is used. Also, it is convenient to think of charge as flowing from positive to negative, just like heat—from hot to cold, or like water—downhill.

As we mentioned in Sect. 3.3, the amount of charge on the positive plate of the capacitor is proportional to the potential drop between the plates: this is *Volta's law of capacitance*, named so after Alessandro Volta who discovered it in 1776. The coefficient of proportionality in Volta's law—the capacitance—depends primarily on the geometry of the plates and the material between them; its nature was extensively investigated in the 1830's by Michael Faraday and therefore the unit of capacitance, *farad*, is named after him.

Faraday's experiments overshadowed Volta's to such an extent that few textbooks bother to call $Q = CV$ Volta's law. Nevertheless, Volta's contribution to the study of capacitors lives on through the word *condenser* which he derived from the Italian word *condensatore*, and which is still the term for capacitor in some languages. Also, the potential difference is called *voltage* and is measured in *volts*.

While the V-notation for voltage is an homage to Volta, the I-notation for electric current is due to André-Marie Ampére for whom the unit of current, *ampere*, is named. In order to avoid confusing the current i with the imaginary unit $\sqrt{-1}$, engineers relabel the latter j.

Kirchhoff announced his current and voltage laws in 1845. The current law is an expression of the law of conservation of charge which was first intuited by Franklin; the voltage law is an expression of the law of conservation of energy. As we noted earlier, Kirchhoff also reformulated Ohm's law in vector form. For more on the early days of electromagnetism we recommend [4].

3.10 Comments and Bibliography

Motion with resistance—the subject of Sect. 3.5—was first modeled, unsurprisingly, by Newton. In the Principia Newton argued that a cannonball of diameter d moving with velocity v will displace air of density ρ at the mass rate

$$\frac{dm}{dt} = \rho \frac{\pi}{4} d^2 v$$

and, since the acceleration of the displaced air is negligible, the rate of change of its momentum is proportional to

$$\frac{dm}{dt} v = \rho \frac{\pi}{4} d^2 v^2,$$

and so is the drag force exerted on the cannonball. Newton's resistance law is thus quadratic in velocity. It is usually written in the form

$$F = C_d \rho \frac{\pi}{8} d^2 v^2, \tag{3.32}$$

where C_d is the drag coefficient.

The applicability of Newton's resistance law (3.32) depends on the value of *Reynolds number* Re which may be thought of as the ratio of inertial forces to viscous forces. For a cannonball flying through air

$$\mathrm{Re} = \frac{\rho d v}{\mu},$$

where μ is air's viscosity. For $\mathrm{Re} > 1 \times 10^3$ Newton's law (3.32) is valid, but for low Reynolds numbers it should be replaced with Stokes law

$$F = 3 \pi \mu d v, \tag{3.33}$$

which is linear in velocity.

Mathematical theories of running and other sports were pioneered by the biologist Hill (see [3, 5, 6]). In [3] the distance travelled by the runner was modeled by Newton's second law which is equivalent to (3.18).

In [7] Keller reformulated Hill's model as a control problem which he solved using the calculus of variations. Keller found optimal running strategies for sprinters and long distance runners and computed bounds for a number of world records. For more on the mathematical theory of running we recommend Keller's original article and the more recent article by Pritchard [13].

The barometric law (3.25), derived in Sect. 3.7, was first empirically established by Sir Edmond Halley in 1686. We found its derivation in the second volume of Thomson and Tate's *Treatise on Natural Philosophy* which was published in 1903, but it must have been known well before that.

NASA divides atmosphere into layers based on inversions of the temperature gradient. A simplified model of air temperature is the piecewise linear function with h measured in meters and T in degrees Celsius:

$$T(h) = \begin{cases} 15.04 - .00649\,h, & 0 \le h \le 11000, \\ -56.46, & 11000 < h \le 25000, \\ -131.21 + .00299\,h, & 25000 < h. \end{cases} \qquad (3.34)$$

The actual 1976 standard atmosphere model is more complicated; for details, see [12].

The electro-thermal analogy is a two-way street, but it is much more relevant in mechanical engineering than in electrical engineering. While it is relatively simple to build an equivalent electric circuit for a thermal circuit, building an equivalent thermal circuit for an electric circuit is often neither simple nor practical. For extension of electro-thermal analogy to conductive heat transfer see [16]; electric analogs for radiative heat transfer are discussed in [9].

3.11 Exercises

1. Bring water to boil, pour it in a cup, and measure its temperature as it cools—this can be done with a kitchen thermometer. Does the temperature obey (3.8)? If not, what are the causes?
2. Equation (3.29) models the temperature of an object that is simultaneously affected by convective and radiative heat transfer. The temperature of a body cooling or heating through radiation alone is modeled by the simpler equation:

$$c\rho V \frac{dT}{dt} = -\epsilon \sigma A \left(T^4 - T_\infty^4\right). \qquad (3.35)$$

   Solve (3.35) using separation of variables and validate the solution numerically. Determine when, if ever, the temperature of the body equilibrates with the temperature of the environment.
3. The cooling light bulb in Sect. 3.1 loses some of its energy through thermal radiation. Use Eq. (3.29) as an improved model for the data in Fig. 3.1 to determine if radiative loss is significant.
4. This problem requires familiarity with electric circuits. Figure 3.9 shows an RC-circuit loaded with a probe of an oscilloscope: R_p and C_p are the resistance and capacitance of the probe, respectively; the voltage source E accounts for soakage. Set up a model for the data in Fig. (3.4) as the voltage across R_p. For Digilent Analog Discovery™ the nominal values are $R_p = 1\,\text{M}\Omega$ and $C_p = 24\,\text{pF}$.

   The same circuit can be used to describe the light bulb from Sect. 3.1 loaded by the temperature probe.
5. In [1] the authors recorded falling balls of various kinds and used machine learning to infer the underlying physics laws from the recordings. Table 3.10 shows the height of one of the wiffle balls used in that study. Test the model (3.18) on this data; for more

3.11 Exercises

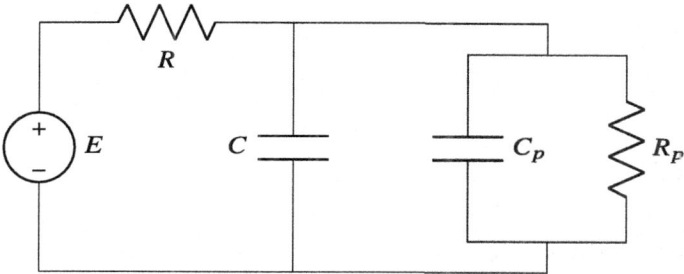

Fig. 3.9 Probe-loaded RC-circuit

t (s)	h (m)	t (s)	h (m)	t (s)	h (m)	t (s)	h (m)
0.00	46.45	0.94	43.16	1.87	34.84	2.81	23.26
0.07	46.48	1.00	42.73	1.94	34.06	2.87	22.33
0.14	46.45	1.07	42.33	2.00	33.21	2.94	21.55
0.20	46.37	1.14	41.82	2.07	32.51	3.01	20.54
0.27	46.22	1.20	41.23	2.14	31.65	3.07	19.53
0.34	46.00	1.27	40.77	2.20	30.88	3.14	18.68
0.40	45.77	1.34	40.20	2.27	30.18	3.21	17.74
0.47	45.60	1.40	39.58	2.34	29.24	3.27	16.65
0.54	45.37	1.47	38.95	2.40	28.47	3.34	15.80
0.60	45.03	1.54	38.30	2.47	27.61	3.41	14.76
0.67	44.72	1.60	37.65	2.54	26.76	3.47	13.79
0.74	44.41	1.67	37.02	2.61	25.90	3.54	12.77
0.80	44.04	1.74	36.31	2.67	24.97	3.61	11.76
0.87	43.64	1.80	35.55	2.74	24.04	3.67	10.96

Fig. 3.10 Ball drop data from [1] (second drop of the yellow wiffle ball)

data and some interesting insights into machine learning consult [1] and the references therein.

6. Modify the derivation of the barometric law in Sect. 3.7 by assuming that the temperature of air is a linear function of altitude: $T = a + bh$. Estimate parameters a and b using the data in Fig. 3.8 and compare the result with (3.34).

References

1. B. de Silva, D.M. Higdon, S.L. Brunton, J.N. Kutz, Discovery of physics from data: Universal laws and discrepancies. Front. Artif. Intell. **3**, (2019)
2. A. Fick Dr., V. on liquid diffusion. The London, Edinburgh, and Dublin Philosophical Magazine and Journal of Science **10**(63), 30–39 (1855)
3. K. Furusawa, Archibald Vivian Hill, and J. L. Parkinson. The dynamics of "sprint" running. Proc. R. Soc. Lond. Ser. B Contain. Pap. Biol. Character **102**(713), 29–42 (1927)
4. J.L. Heilbron, *Electricity in the 17th & 18th Centuries: A Study in Early Modern Physics* (Dover Books on Physics, Dover Publications, 1999)
5. A.V. Hill, The air-resistance to a runner. Proc. R. Soc. Lond. Ser. B Contain. Pap. Biol. Character **102**(718), 380–385 (1928)
6. A.V. Hill, The physiological basis of athletic records. Lancet **206**(5323), 481–486 (1925). Originally published as Volume 2, Issue 5323
7. J.B. Keller, A theory of competitive running. Phys. Today **26**, 43–47 (1973)
8. S. Maruyama, S. Moriya, Newton's law of cooling: follow up and exploration. Int. J. Heat Mass Transf. **164**(01), 120544 (2021)
9. S. Maslovski, C. Simovski, S. Tretyakov, Equivalent circuit model of radiative heat transfer. Phys. Rev. B **87**(04), 155124 (2013)
10. J.C. Maxwell, *A Treatise on Electricity and Magnetism*, vol. 1 (Dover Books on Physics, Dover Publications, 1954)
11. G.S. Ohm, W. Francis, T.D. Lockwood, *The Galvanic Circuit Investigated Mathematically*. Van Nostrand's Science Series (D. Van Nostrand Company, 1891)
12. United States Committee on Extension to the Standard Atmosphere. *U.S. Standard Atmosphere, 1976*. NOAA-SIT 76-1562. National Oceanic and Atmospheric Administration (1976)
13. W.G. Pritchard, Mathematical models of running. SIAM Rev. **35**(3), 359–379 (1993)
14. J. Ruffner, Reinterpretation of the Genesis of Newton's "Law of Cooling." Arch. Hist. Exact Sci. **2**, 138–152 (1963)
15. D.L. Simms, Newton's contribution to the science of heat. Ann. Sci. **61**(1), 33–77 (2004)
16. R.C. Steere, Solution of two dimensional transient heat flow problems by electrical analogue. Phys. Educ. **6**(6), 443–447 (1971)
17. R. Winterton, Early study of heat transfer: Newton and Fourier. Heat Transf. Eng. **22**, 03–11 (2001)

Memoryless Processes

4

In Chap. 3 we saw how exponential laws originate as solutions of linearized ODE. In this chapter we will examine another source of exponential laws: memoryless processes.

A classic example of a memoryless process is radioactive decay. Let $N = N(t)$ be the number of unstable atoms in a radioactive sample. In textbooks the formula

$$N = N_0 e^{-kt} \tag{4.1}$$

is often "derived" by solving the IVP

$$\frac{dN}{dt} = -kN, \quad N(0) = N_0, \tag{4.2}$$

which is either postulated or presented as an expression of the law of mass action. However, (4.1) is not the true number of radioactive atoms. In actuality, N is an integer-valued random variable whose plot looks like the staircase in Fig. 4.1.

In the next section we will derive the distribution of N and show that (4.1) is its mean. Then we will discuss two other examples of memoryless processes: absorption of light and mutarotation of glucose.

4.1 Radioactive Decay as a Memoryless Process

The probabilistic model of radioactive decay is based on two assumptions:

1. Atoms have no memory of their past history.
2. Atoms are indistinguishable from and decay independently of each other.

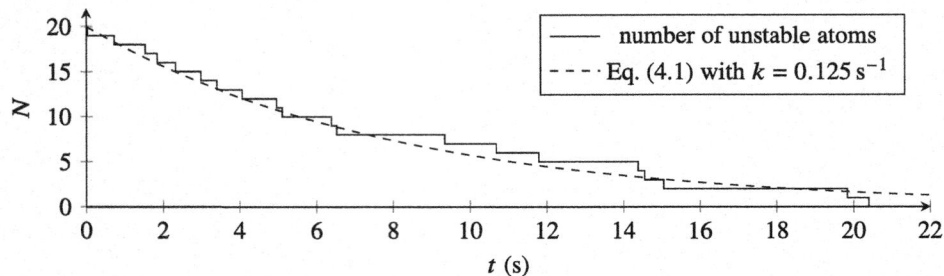

Fig. 4.1 Radioactive decay of a sample consisting of 20 atoms

Let τ be the disintegration time of an atom and let $p(t)$ be the probability of an atom surviving—that is, not disintegrating—for t seconds: $p(t) = P(\tau > t)$. By Assumption 2, the probability of k out of N_0 atoms surviving at time t is

$$P(N(t) = k) = \binom{N_0}{k}(p(t))^k (1 - p(t))^{N_0 - k}. \tag{4.3}$$

To find $p(t)$, consider $p(t + \Delta t)$ as probability of a compound event: the survival of t seconds (with probability $p(t)$) followed by the survival of additional Δt seconds. By Assumption 1, the survival of additional Δt seconds is independent of the prior survival of t seconds and has probability $p(\Delta t)$. Therefore,

$$p(t + \Delta t) = p(t)\, p(\Delta t),$$

which we can rewrite, suggestively, as

$$\frac{p(t + \Delta t) - p(t)}{\Delta t} = -\frac{1 - p(\Delta t)}{\Delta t} p(t).$$

Taking the limit, as $\Delta t \to 0^+$, results in the ODE

$$\frac{dp}{dt} = -k\, p, \quad k = \lim_{\Delta t \to 0^+} \frac{1 - p(\Delta t)}{\Delta t} > 0,$$

whose solution with $p(0) = 1$—surviving zero seconds is a sure event—is

$$p(t) = e^{-kt}. \tag{4.4}$$

Together, Eqs. (4.3) and (4.4) form the true law of exponential decay.

The mean of (4.3) is $\mu(t) = N_0\, p(t) = N_0\, e^{-kt}$. We will now show, using simulated data, that for large N_0 it is close to the actual count.

The distribution of disintegration times

4.1 Radioactive Decay as a Memoryless Process

$$F_\tau(t) = P(\tau \le t) = 1 - P(\tau > t) = 1 - p(t) = 1 - e^{-kt}$$

is the same as the distribution of $x = -k^{-1} \ln(1-u)$ where u is uniform on [0, 1]. Indeed,

$$F_x(t) = P\left(-k^{-1} \ln(1-u) \le t\right) = P\left(u \le 1 - e^{-kt}\right) = 1 - e^{-kt}.$$

This means that we can sample F_τ using MATLAB's uniform random number generator rand.

Figure 4.2 illustrates radioactive decay of 500 atoms with $k = 0.125\,\mathrm{s}^{-1}$ using data generated by the following code:

```
k    = .125;  N0 = 500;
tau  = [0;  sort(-(1/k)*log(1-rand(N0,1)))];
N    = (N0:-1:0)';
```

On the left we plotted the staircase count and (4.1); on the right we compared the histogram of decay times with their probability density function

$$\frac{dF_\tau}{dt} = k e^{-kt}.$$

Notice how close the staircase plot is to the plot of (4.1).

As N_0 is increased, the stochasticity in Fig. 4.2 becomes less and less apparent. With $N_0 = 1 \times 10^6$ the true count N cannot be visually distinguished from (4.1). For comparison, the number of atoms in one mole is $6.02214076 \times 10^{23}$ (the Avogadro constant). Even trace amounts of radioactive substances have trillions of atoms.

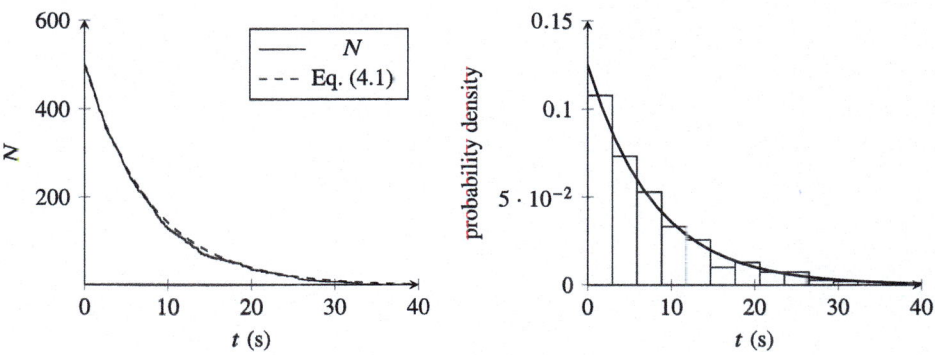

Fig. 4.2 Radioactive decay of a sample consisting of 500 atoms

4.2 Markov Description of Radioactive Decay

Suppose that we observe an unstable atom X at times

$$t_n = T\frac{n}{N}, \quad n = 0, \ldots, N.$$

During each time step the atom may spontaneously decay into a more stable form Y. As we showed in Sect. 4.1, the probability of not decaying—surviving Δt seconds is $p(\Delta t) = e^{-k\Delta t}$. The probability of the transition $X \to Y$ is then $1 - e^{-k\Delta t}$.

We can think of the atom abstractly as a system with two states: X and Y. At each time step the atom either stays in state X with probability $e^{-k\Delta t}$ or transitions into state Y with probability $1 - e^{-k\Delta t}$. Once the atom transitions into state Y, it stays in that state; Y is an *absorbing state*, as indicated by its loop having transition probability 1.

Let $x_n = x(t_n)$ and $y_n = y(t_n)$ denote the probabilities of finding the atom at time t_n in states X and Y, respectively. According to the transitions shown in Fig. 4.3,

$$x_{n+1} = e^{-k\Delta t} x_n, \quad y_{n+1} = (1 - e^{-k\Delta t}) x_n + y_n.$$

If we write these equations in matrix form

$$\begin{bmatrix} x_{n+1} \\ y_{n+1} \end{bmatrix} = \begin{bmatrix} e^{-k\Delta t} & 0 \\ 1 - e^{-k\Delta t} & 1 \end{bmatrix} \begin{bmatrix} x_n \\ y_n \end{bmatrix} \tag{4.5}$$

then

$$\begin{bmatrix} x_n \\ y_n \end{bmatrix} = \begin{bmatrix} e^{-k\Delta t} & 0 \\ 1 - e^{-k\Delta t} & 1 \end{bmatrix}^n \begin{bmatrix} x_0 \\ y_0 \end{bmatrix} = \begin{bmatrix} e^{-kn\Delta t} & 0 \\ 1 - e^{-kn\Delta t} & 1 \end{bmatrix} \begin{bmatrix} x_0 \\ y_0 \end{bmatrix}. \tag{4.6}$$

Here the *transition probability matrix* is simple enough for its n-th power to be guessed from a pattern (exercise). According to (4.6), if the atom is initially in state X then at time $T = N\Delta t$ the probability that it is still in state X is given by $x_N = e^{-kN\Delta t} = e^{-kT}$: this is consistent with the formula for the survival probability $p(t)$ derived in Sect. 4.1.

Equation (4.5) is equivalent to

Fig. 4.3 Markov chain for radioactive decay

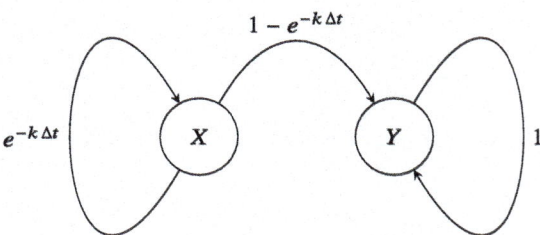

$$\frac{1}{\Delta t}\left(\begin{bmatrix} x_{n+1} \\ y_{n+1} \end{bmatrix} - \begin{bmatrix} x_n \\ y_n \end{bmatrix}\right) = \frac{1}{\Delta t}\begin{bmatrix} e^{-k\Delta t} - 1 & 0 \\ 1 - e^{-k\Delta t} & 0 \end{bmatrix}\begin{bmatrix} x_n \\ y_n \end{bmatrix}. \tag{4.7}$$

If we simultaneously let $\Delta t \to 0^+$ and $n \to \infty$ so that $n\,\Delta t = t$ remains fixed, Eq. (4.7) becomes

$$\frac{d}{dt}\begin{bmatrix} x \\ y \end{bmatrix} = \begin{bmatrix} -k & 0 \\ k & 0 \end{bmatrix}\begin{bmatrix} x \\ y \end{bmatrix} \tag{4.8}$$

This is *Kolmogorov's forward equation* for continuous time Markov process modeling radioactive decay. If instead of regarding x and y as probabilities of a single atom being in states X and Y we regard them as concentrations of the species X and Y, Eq. (4.8) becomes an expression of the law of mass action for the reaction $X \xrightarrow{k} Y$. Writing (4.8) as a system

$$\frac{dx}{dt} = -kx, \quad \frac{dy}{dt} = kx$$

shows that $x = x_0 e^{-kt}$ and $y = y_0 + x_0 - x_0 e^{-kt}$. In Chap. 5 we will derive the solution of (4.8) in matrix form, without splitting it into a system of equations.

4.3 Absorption of Light

The table in Fig. 4.4 shows how the intensity of light emitted by 650 nm laser is diminished by a layer of aqueous solution of paint pigment "Cadmium Red"; the data was acquired with PASCO™ CI-6504A light sensor.

As photons emitted by the laser propagate through the solution, they get randomly scattered and absorbed in much the same way as radioactive atoms randomly decay. Absorption of light is thus a memoryless process where the role of time is played by distance traveled through absorbing medium.

The plot of the data in the upper right corner of Fig. 4.4 shows that the first measurement (filled circle) is an outlier. We discarded it and fitted *Lambert law*

$$I(x) = I_0 e^{-kx} \tag{4.9}$$

to the rest of the data using the same methodology as in Chap. 3. The resulting parameter estimates are

$$I_0 = 1.521843845423095 \times 10^2 \text{ lx}, \quad k = 0.153613239527374 \text{ cm}^{-1}.$$

The standard deviation is $\hat{\sigma} = 0.536130576409234$ lx. Since our data set is small and we do not know the accuracy of the CI-6504A light sensor, we cannot perform meaningful statistical analysis of the residual. Still, the exponential fit in Fig. 4.4 leaves no doubts about the validity of (4.9).

x (cm)	I (lx)
2	95.5
3	94.7
4	82.8
5	70.9
6	61.6
7	52.9
8	44.8
9	37.4
10	32.5
11	27.9
12	24.1
13	20.6
14	17.8
15	14.6
16	12.7
17	10.9
18	9.4
19	8.1
20	6.9
21	6.1

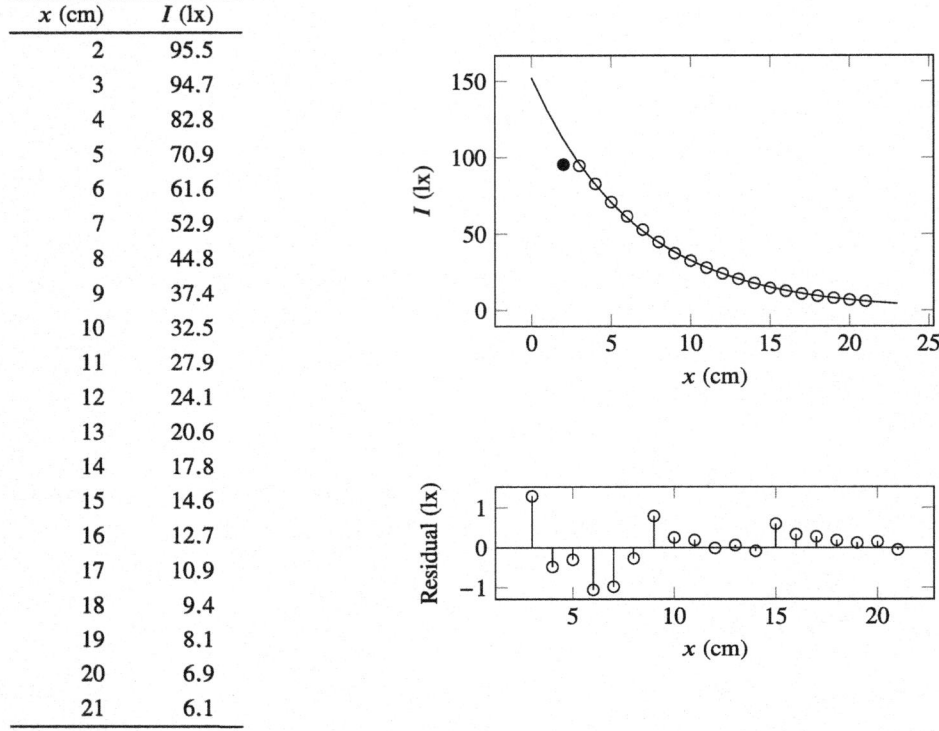

Fig. 4.4 Absorption of light

4.4 Mutarotation of Glucose

Glucose exists in two forms which, in solution, transform into one another:

$$\alpha\text{-Glucose} \underset{k_2}{\overset{k_1}{\rightleftharpoons}} \beta\text{-Glucose} \tag{4.10}$$

What happens during the transformation is more than a spontaneous rearrangement of one of the OH groups. However we will ignore the subtleties of the chemical mechanism and identify (4.10) with the Markov chain in Fig. 4.5.

The interpretation of Fig. 4.5 is similar to that of Fig. 4.3. Imagine observing a molecule of glucose at intervals of Δt seconds. During an interval α-glucose can "survive" with

4.4 Mutarotation of Glucose

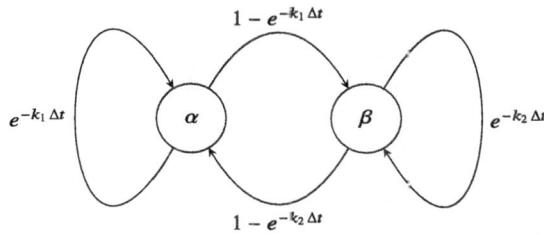

Fig. 4.5 Markov chain for the transformation between α- and β-forms of glucose

probability $e^{-k_1 \Delta t}$ or "decay" into β-form with complimentary probability $1 - e^{-k_1 \Delta t}$; the same holds for β-glucose but with the rate constant k_2.

Let $x_n = x(t_n)$ and $y_n = y(t_n)$ denote the probabilities of finding the glucose molecule at time t_n in states α and β, respectively. The analog of (4.5) is

$$\begin{bmatrix} x_{n+1} \\ y_{n+1} \end{bmatrix} = \begin{bmatrix} e^{-k_1 \Delta t} & 1 - e^{-k_2 \Delta t} \\ 1 - e^{-k_1 \Delta t} & e^{-k_2 \Delta t} \end{bmatrix} \begin{bmatrix} x_n \\ y_n \end{bmatrix} \quad (4.11)$$

and we can write

$$\begin{bmatrix} x_n \\ y_n \end{bmatrix} = \begin{bmatrix} e^{-k_1 \Delta t} & 1 - e^{-k_2 \Delta t} \\ 1 - e^{-k_1 \Delta t} & e^{-k_2 \Delta t} \end{bmatrix}^n \begin{bmatrix} x_0 \\ y_0 \end{bmatrix}. \quad (4.12)$$

However, powering the matrix of transition probabilities in (4.12) requires *eigendecomposition* which will be discussed in Chap. 7. In the meantime, mimicking the development in Sect. 4.2, we can rewrite (4.11) as

$$\frac{1}{\Delta t} \left(\begin{bmatrix} x_{n+1} \\ y_{n+1} \end{bmatrix} - \begin{bmatrix} x_n \\ y_n \end{bmatrix} \right) = \frac{1}{\Delta t} \begin{bmatrix} e^{-k_1 \Delta t} - 1 & 1 - e^{-k_2 \Delta t} \\ 1 - e^{-k_1 \Delta t} & e^{-k_2 \Delta t} - 1 \end{bmatrix} \begin{bmatrix} x_n \\ y_n \end{bmatrix}$$

Letting $\Delta t \to 0^+$ and $n \to \infty$ so that $n \Delta t = t$ gives Kolmogorov's forward equation:

$$\frac{d}{dt} \begin{bmatrix} x \\ y \end{bmatrix} = \begin{bmatrix} -k_1 & k_2 \\ k_1 & -k_2 \end{bmatrix} \begin{bmatrix} x \\ y \end{bmatrix}. \quad (4.13)$$

Writing (4.13) as a system of equations

$$\frac{dx}{dt} = -k_1 x + k_2 y, \quad \frac{dy}{dt} = k_1 x - k_2 y,$$

shows that $dx/dt = -dy/dt$ and, therefore, $x - x_0 = y_0 - y$. Elimination of y from the system gives a separable ODE for x

$$\frac{dx}{dt} = -k_1 x + k_2 (x_0 + y_0 - x)$$

with the solution:

$$x = \frac{k_1}{k_1+k_2}\left(\frac{k_2}{k_1}+e^{-(k_1+k_2)t}\right)x_0 + \frac{k_2}{k_1+k_2}\left(1-e^{-(k_1+k_2)t}\right)y_0. \qquad (4.14)$$

Once x is computed using (4.14), y can be found as: $y = x_0 + y_0 - x$. In the context of the Markov model, x and y are probabilities of finding a glucose molecule at time t in state α and β, respectively. However, since the number of glucose molecules in a typical reaction is comparable to Avogadro number, we can regard x and y as concentrations of α- and β-glucose, as we will do from now on.

Glucose is an optically active substance: its solution rotates polarized light. Both α- and β-forms rotate light in the same direction but by different amounts. If a solution of pure α-glucose is placed in a tube of a polarimeter then, as α-glucose turns into β-glucose, the measured angle of rotation will gradually decrease until equilibrium between the forms is reached. This process is called *mutarotation*.

Figure 4.6 shows the difference between the angle of rotation α_t and its limiting value α_∞ for the solution of initially pure α-glucose with concentration 0.3193 g.mol./l.; the data (taken from [2]) was collected at 278.26 °K with a polarimeter having 4 dm optical path. The second column shows observed values of $\Delta\alpha = \alpha_t - \alpha_\infty$; the third column shows values calculated using $\Delta\alpha = \Delta\alpha_0\, e^{-kt}$.

The exponential law for $\Delta\alpha$ follows from (4.14) and *Biot's law*. According to the latter, the observed angle of rotation is $\alpha_t = l\left([\alpha]_x\, x + [\alpha]_y\, y\right)$ where x and y are the concentrations of the α- and β-forms; $[\alpha]_x$ and $[\alpha]_y$ are their specific rotations; and l is the length of the optical path. As follows from (4.14), for a solution of pure α-glucose with initial concentration x_0

$$x = x_0 \frac{k_1}{k_1+k_2}\left(\frac{k_2}{k_1}+e^{-(k_1+k_2)t}\right), \quad y = x_0 - x.$$

Therefore
$$\alpha_t = \alpha_\infty + (\alpha_0 - \alpha_\infty)\, e^{-(k_1+k_2)t},$$

where
$$\alpha_0 = x_0\, l\, [\alpha]_x, \quad \alpha_\infty = x_0\, l\, \frac{k_1\,[\alpha]_y + k_2\,[\alpha]_x}{k_1+k_2}.$$

This is equivalent to $\Delta\alpha = \Delta\alpha_0\, e^{-kt}$ with $k = k_1 + k_2$.

After performing nonlinear least squares—a computation that would have been prohibitively expensive in 1940 when the data in Fig. 4.6 was collected—we found

$$\widehat{\Delta\alpha_0} = 13.030116464179942°, \quad \widehat{k} = 5.146016132200587 \times 10^{-5}\,\text{s}^{-1}.$$

If we accept $\widehat{k}$ as the true value, then the value $k = 5.13 \times 10^{-5}\,\text{s}^{-1}$ reported in [2] has relative error of only about 3%; the relative errors in the calculated values of $\Delta\alpha$ shown in Fig. 4.6 do not exceed 5%.

t (min)	Δα° obs.	Δα° calc.
20	12.23	12.26
30	11.86	11.88
40	11.50	11.52
50	11.17	11.17
60	10.84	10.83
75	10.35	10.35
90	9.88	9.88
105	9.43	9.43
125	8.87	8.87
154	8.10	8.10
161	7.90	7.94
180	7.49	7.49
190	7.27	7.26
210	6.82	6.83
220	6.61	6.62
240	6.22	6.23
260	5.83	5.85
280	5.49	5.50
300	5.16	5.17
330	4.72	4.72
360	4.28	4.30
390	3.90	3.92
420	3.55	3.58
450	3.23	3.26
480	2.94	2.97

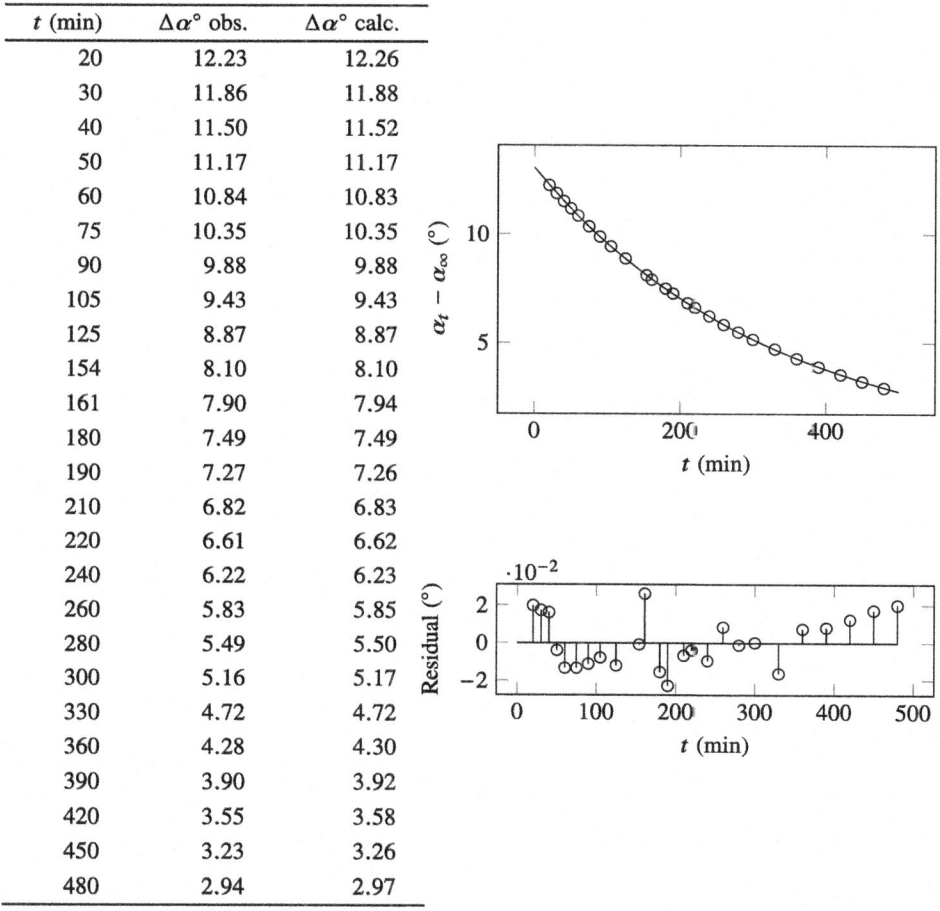

Fig. 4.6 Mutarotation of α-glucose (from [2])

We also found $\hat{\sigma} = 0.013540802717565°$. This is considerably greater than the 0.002° accuracy of the polarimeter used in [2] but is consistent with the reproducibility of readings to 0.01° mentioned by the authors.

4.5 Comments and Bibliography

A thorough review of stochastic modeling in chemical kinetics can be found in [4]. Unfortunately, as pointed out in [1], chemical kinetics is often taught as a purely experimental discipline. As a result, many students of natural sciences are unaware of probabilistic foundations of chemical kinetics and think that the law of mass action is as empirical today as it was when discovered by Guldberg and Waage in the late 19th century.

In Sect. 4.3 we referenced (4.9) as Lambert law. For more on its history and the contributions of Bouguer and Beer see [3].

For simplicity, we labeled the molecules in (4.10) as α- and β-glucose: a more accurate labeling would be α- and β-D-glucopyranose. In solution the cyclical form of glycose—glucopyranose—first transforms into linear form. That linear form can re-close as either α- or β- cyclical form. The β-form is thermodynamically more stable. Therefore, at equilibrium, the room temperature solution of D-glucose has α- and β-forms in approximate ratio of $1 : 2$.

4.6 Exercises

1. Compute the standard deviation σ of (4.3) and compare it with the mean μ: for what initial number of unstable atoms N_0 is σ less than $1 \times 10^{-6} \mu$? How does that number compare to the Avogadro constant?
2. In 1949 Willard Libby discovered a method of dating archeological finds based on their carbon-14 content. ^{14}C is an unstable isotope of carbon created in the Earth's atmosphere by cosmic rays and incorporated into carbon dioxide. Living organisms ingest it at the rate that exactly balances its decay. However, after an organism dies, it ceases to replenish its supply of ^{14}C which then decays exponentially. Let $x(t)$ denote the ^{14}C content of biological remains t years after death and let $r(t) = -dx/dt$ be the rate of disintegration: this is the data that radioactivity detectors measure. In 1950 a sample of charcoal from Lascaux Cave in France gave an average count of $r = 0.97$ disintegrations per minute per gram. Estimate the date of occupation and hence the probable age of the paintings in the Lascaux Cave using 6.68 as the average count of disintegrations per minute per gram for living wood.
3. Analysis of Markov chains requires computation of powers of matrices. We will explain how to power matrices using *eigendecomposition* in Chap. 7. In preparation, find the nth power of each of the following matrices by following the patterns:

$$\begin{bmatrix} a & 0 \\ 0 & b \end{bmatrix}, \begin{bmatrix} 0 & a \\ b & 0 \end{bmatrix}, \begin{bmatrix} a & 0 \\ b & 1 \end{bmatrix}.$$

The patterns can be easily observed in MATLAB.

References

1. S.P. Huestis, Understanding the origin and meaning of the radioactive decay equation. J. Geosci. Educ. **50**(5), 524–527 (2002)
2. J.C. Kendrew, E.A. Moelwyn-Hughes, The Kinetics of Mutarotation in Solution. Proc. R. Soc. Lond. Ser. A **176**(966), 352–367 (1940)
3. T. Mayerhöfer, S. Pahlow, J. Popp, The Bouguer-Beer-Lambert law: shining light on the obscure. ChemPhysChem **21**, 07 (2020)
4. D.A. McQuarrie, Stochastic approach to chemical kinetics. J. Appl. Probab. **4**(3), 413–478 (1967)

Exponential Function 5

In Chaps. 3 and 4 we showed how exponential laws arise as solutions of simple linear ODE. As we will see in Chap. 7, the solutions of all linear ODE with constant coefficients involve exponentials in one way or another. In preparation for that, we need to rigorously define the exponential function and extend it into the complex and matrix domains.

5.1 Calculus Definition

Euler introduced exponential function as the limit

$$e^x = \lim_{n \to \infty} \left(1 + \frac{x}{n}\right)^n. \tag{5.1}$$

He then used the binomial expansion

$$\left(1 + \frac{x}{n}\right)^n = 1 + \binom{n}{1}\frac{x}{n} + \binom{n}{2}\left(\frac{x}{n}\right)^2 + \binom{n}{3}\left(\frac{x}{n}\right)^3 + \cdots + \binom{n}{n}\left(\frac{x}{n}\right)^n$$

and the fact that for fixed k

$$\lim_{n \to \infty} \binom{n}{k}\left(\frac{x}{n}\right)^k = \frac{x^k}{k!}$$

to derive the power series

$$e^x = \sum_{k=0}^{\infty} \frac{x^k}{k!} = 1 + x + \frac{x^2}{2} + \frac{x^3}{6} + \cdots \tag{5.2}$$

which became the standard Calculus definition.

Equation (5.2) is a convenient definition for a number of reasons. Firstly, it is computable—by truncating the series we can approximate e^x with a polynomial. Secondly, it readily yields various properties of the exponential function: the term-by-term differentiation

$$\frac{d}{dx}\left(1+x+\frac{x^2}{2}+\frac{x^3}{6}+\frac{x^4}{24}+\cdots\right) = 1+x+\frac{x^2}{2}+\frac{x^3}{6}+\cdots$$

shows that the derivative of e^x is itself; the multiplication of series

$$\left(1+x+\frac{x^2}{2}+\frac{x^3}{6}+\cdots\right)\left(1+y+\frac{y^2}{2}+\frac{y^3}{6}+\cdots\right) = 1$$
$$+ (x+y) + \left(\frac{x^2}{2}+xy+\frac{y^2}{2}\right) + \left(\frac{x^3}{6}+\frac{x^2}{2}y+x\frac{y^2}{2}+\frac{y^3}{6}\right) + \cdots$$
$$= 1 + (x+y) + \frac{(x+y)^2}{2} + \frac{(x+y)^3}{6} + \cdots$$

proves the formula

$$e^x e^y = e^{x+y}, \tag{5.3}$$

and so on. Thirdly, Eq. (5.2) makes sense for complex numbers. Indeed, the series

$$\sum_{k=0}^{\infty} \frac{z^k}{k!} = \sum_{k=0}^{\infty} \frac{(x+yi)^k}{k!} \tag{5.4}$$

is absolutely convergent:

$$\left|\sum_{k=0}^{\infty} \frac{z^k}{k!}\right| \leq \sum_{k=0}^{\infty} \frac{|z|^k}{k!} = e^{|z|}.$$

Therefore (5.4) defines a complex function which, for obvious reasons, is denoted e^z.

5.2 Complex Exponentials

If we stipulate that the addition formula (5.3) should hold for complex numbers, then

$$e^z = e^{x+yi} = e^x e^{yi}.$$

To compute e^{yi}, use (5.2) and split the sum into even and odd terms:

$$e^{yi} = \sum_{k=0}^{\infty} \frac{(yi)^k}{k!} = \sum_{k=0}^{\infty} \frac{(yi)^{2k}}{(2k)!} + \sum_{k=0}^{\infty} \frac{(yi)^{2k+1}}{(2k+1)!}$$
$$= \sum_{k=0}^{\infty} (-1)^k \frac{y^{2k}}{(2k)!} + i \sum_{k=0}^{\infty} (-1)^k \frac{y^{2k+1}}{(2k+1)!}$$
$$= \cos(y) + i \sin(y).$$

5.2 Complex Exponentials

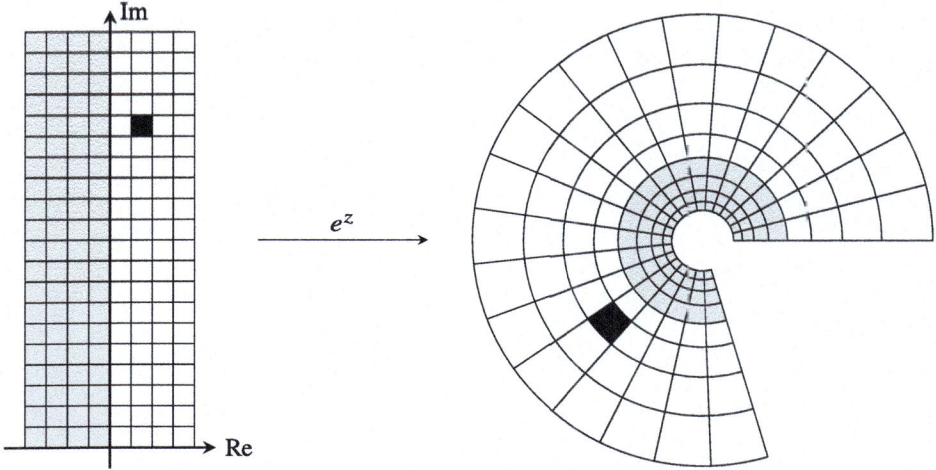

Fig. 5.1 Conformal plot of e^z

It follows that
$$e^{x+yi} = e^x \left(\cos(y) + i\,\sin(y)\right). \tag{5.5}$$

We will refer to Eq. (5.5) as *Euler's formula*.

Figure 5.1 shows the *conformal plot* of $w = e^z$.

To produce Fig. 5.1 we covered the rectangle
$$\{z \in \mathbb{C} \mid -1 \leq \mathrm{Re}(z) \leq 1,\ 0 \leq \mathrm{Im}(z) \leq 5\}$$

with a Cartesian grid, as shown on the left, and plotted the image of that grid under e^z, as shown on the right. As follows from Euler's formula (5.5), e^z maps vertical lines into circular arcs and horizontal lines into rays issuing from the origin: Cartesian grid is thus transformed into polar grid.

In complex analysis it is shown that if γ_1 and γ_2 are two curves in the z-plane intersecting at angle α then their images e^{γ_1} and e^{γ_2} in the w-plane intersect at the same angle α. Angle-preserving transformations are called *conformal*: that is why Fig. 5.1 is a conformal plot.

5.3 Complex Trigonometry

Using Euler's formula (5.5), one can express cosine and sine in terms of complex exponentials:
$$\cos(\omega t) = \frac{e^{i\omega t} + e^{-i\omega t}}{2}, \quad \sin(\omega t) = \frac{e^{i\omega t} - e^{-i\omega t}}{2i}. \tag{5.6}$$

Equation (5.6) profoundly simplifies many computations involving trigonometric functions. For instance, the *Laplace transform* of $\cos(\omega t)$ is the following integral:

$$L(\cos(\omega t)) = \int_0^\infty e^{-st} \cos(\omega t) \, dt. \tag{5.7}$$

It can be computed using integration by parts, but a much simpler approach is to use (5.6):

$$\begin{aligned}
\int_0^\infty e^{-st} \cos(\omega t) \, dt &= \int_0^\infty e^{-st} \left(\frac{e^{i\omega t} + e^{-i\omega t}}{2} \right) dt \\
&= \frac{1}{2} \left(\int_0^\infty e^{(-s+i\omega)t} \, dt + \int_0^\infty e^{(-s-i\omega)t} \, dt \right) \\
&= \frac{1}{2} \left(\frac{e^{(-s+i\omega)t}}{-s+i\omega} \bigg|_0^\infty + \frac{e^{(-s-i\omega)t}}{-s-i\omega} \bigg|_0^\infty \right) = \frac{s}{s^2 + \omega^2}.
\end{aligned}$$

In the above computation we treated s as a complex number with negative real part; if $\mathrm{Re}(s) \geq 0$ the integral diverges.

5.4 Matrix Exponentials

In Chap. 4 we encountered matrix-vector ODE while discussing radioactive decay and mutarotation of glucose. We will now show that solutions of such ODE can be expressed as matrix exponentials.

Consider

$$\frac{d\mathbf{u}}{dt} = A\mathbf{u}, \quad \mathbf{u}(0) = \mathbf{u}_0 \tag{5.8}$$

where $\mathbf{u} \in \mathbb{R}^n$ and A is an n-by-n matrix. Let us seek the solution of (5.8) as the power series

$$\mathbf{u} = \mathbf{c}_0 + t\,\mathbf{c}_1 + t^2\,\mathbf{c}_2 + t^3\,\mathbf{c}_3 + \cdots \tag{5.9}$$

where the vector coefficients $\mathbf{c}_k$ are to be determined. Substituting (5.9) into (5.8)

$$\mathbf{c}_1 + 2t\,\mathbf{c}_2 + 3t^2\,\mathbf{c}_3 + \cdots = A\,\mathbf{c}_0 + t\,A\,\mathbf{c}_1 + t^2\,A\,\mathbf{c}_2 + \cdots$$

and equating like coefficients, we get

$$\mathbf{c}_1 = A\,\mathbf{c}_0, \quad 2\,\mathbf{c}_2 = A\,\mathbf{c}_1, \quad 3\,\mathbf{c}_3 = A\,\mathbf{c}_2, \ldots$$

which can be stated as the recurrence relation

$$k\,\mathbf{c}_k = A\,\mathbf{c}_{k-1}, \quad k = 1, 2, 3, \ldots$$

5.4 Matrix Exponentials

Evaluating (5.9) at $t = 0$ shows that $c_0 = u_0$. We can now use the recurrence relation to successively compute

$$c_1 = A\,c_0 = A\,u_0, \quad c_2 = \frac{1}{2} A\,c_1 = \frac{1}{2} A^2\,u_0, \quad c_3 = \frac{1}{3} A\,c_2 = \frac{1}{6} A^3\,u_0, \ldots$$

and, in general,

$$c_k = \frac{1}{k!} A^k\,u_0.$$

The solution of (5.8) is therefore

$$\begin{aligned} u &= u_0 + t\,A\,u_0 + \frac{t^2}{2} A^2\,u_0 + \frac{t^3}{6} A^3\,u_0 + \cdots \\ &= \left(I + t\,A + \frac{(t\,A)^2}{2} + \frac{(t\,A)^3}{6} + \cdots \right) u_0, \end{aligned}$$

where I is the identity matrix. The series in parenthesis is the series (5.2) with $x = t\,A$ and the constant term 1 replaced by its matrix analog I. If we define the matrix exponential e^X as the series

$$e^X = \sum_{k=0}^{\infty} \frac{1}{k!} X^k = I + X + \frac{X^2}{2} + \frac{X^3}{6} + \cdots \tag{5.10}$$

then the solution of (5.8) is simply

$$u = e^{t\,A}\,u_0. \tag{5.11}$$

As an illustration, let us set

$$A = \begin{bmatrix} -k & 0 \\ k & 0 \end{bmatrix}$$

in (5.11). Applying (5.10) (with a different index of summation, so as not to confuse it with the k in the matrix), we get:

$$\begin{aligned} e^{t\,A} &= \sum_{n=0}^{\infty} \frac{1}{n!} \left(t \begin{bmatrix} -k & 0 \\ k & 0 \end{bmatrix} \right)^n = \sum_{n=0}^{\infty} \frac{(k\,t)^n}{n!} \begin{bmatrix} -1 & 0 \\ 1 & 0 \end{bmatrix}^n \\ &= \begin{bmatrix} 1 & 0 \\ 0 & 1 \end{bmatrix} + \sum_{n=1}^{\infty} \frac{(k\,t)^n}{n!} (-1)^{n-1} \begin{bmatrix} -1 & 0 \\ 1 & 0 \end{bmatrix} \\ &= \begin{bmatrix} 1 & 0 \\ 0 & 1 \end{bmatrix} - \left(e^{-k\,t} - 1 \right) \begin{bmatrix} -1 & 0 \\ 1 & 0 \end{bmatrix} = \begin{bmatrix} e^{-k\,t} & 0 \\ 1 - e^{-t} & 1 \end{bmatrix}. \end{aligned}$$

Therefore the solution of

$$\frac{d}{dt} \begin{bmatrix} x \\ y \end{bmatrix} = \begin{bmatrix} -k & 0 \\ k & 0 \end{bmatrix} \begin{bmatrix} x \\ y \end{bmatrix}, \quad \begin{bmatrix} x(0) \\ y(0) \end{bmatrix} = \begin{bmatrix} x_C \\ y_C \end{bmatrix}$$

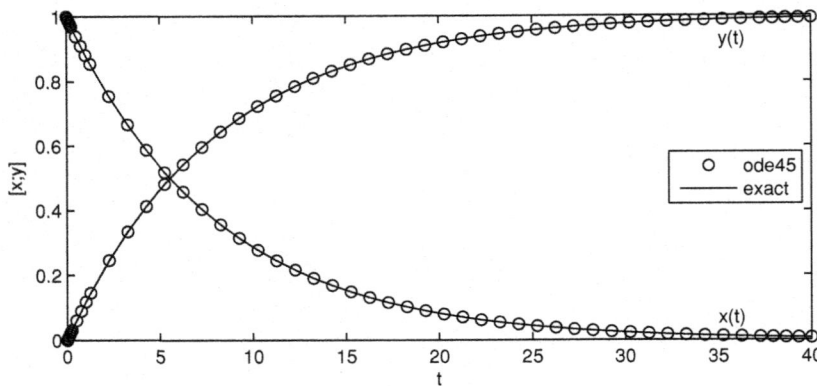

Fig. 5.2 Numerical validation of (5.12) with $k = 0.125$

is
$$\begin{bmatrix} x \\ y \end{bmatrix} = \begin{bmatrix} e^{-kt} & 0 \\ 1 - e^{-t} & 1 \end{bmatrix} \begin{bmatrix} x_0 \\ y_0 \end{bmatrix} \tag{5.12}$$

which matches the solution of (4.8) found in Sect. 4.2.

The following code produces Figure 5.2 which validates (5.12).

```
k = .125; A = [-k 0; k 0];
odefun = @(t,u) A*u; u0 = [1;0];
[t,u] = ode45(odefun,[0 40],u0);
figure; p1 = plot(t,u,'ko'); hold on;
v = zeros(size(u));
for n=1:length(t)
    v(n,:) = [exp(-k*t(n)) 0; 1-exp(-k*t(n)) 1]*u0;
end
p2 = plot(t,v,'k-'); xlabel('t'); ylabel('[x;y]');
legend([p1(1),p2(1)],'ode45','exact','location','east')
text(35,.07,'x(t)'); text(35,.93,'y(t)')
```

In MATLAB matrix exponentials can be computed using the command expm. For instance, the line inside the `for` loop in the above code can be replaced with

```
v(n,:) = expm(t(n)*A)*u0;
```

While matrix exponentials are defined as Taylor series, using (5.10), they are computed using eigendecomposition which is the subject of Sect. 7.1.

5.5 Comments and Bibliography

In pre-Calculus the exponential function $y = a^x$ is defined algebraically for rational inputs

$$a^{\frac{m}{n}} = n\text{th root of } \underbrace{a \times a \times \cdots \times a}_{m \text{ times}}.$$

This creates a deep-seated misconception that

$$e^x = \underbrace{e \times e \times \cdots \times e}_{x \text{ times}}$$

which makes transition to complex exponentials difficult: how do you multiply e by itself complex number of times? This misconception should be dispelled in Calculus with Taylor series. Unfortunately, the subtle point that Taylor series actually define transcendental functions like e^x does not always get across.

Instead of using (5.1) or (5.2), one can define exponential function $y = e^x$ for real inputs in the following ways:

1. As the inverse of the natural logarithm which may be defined as the area under the hyperbola $y = 1/x$.
2. As the continuous solution of the functional equation $f(x + y) = f(x) f(y)$ with $f(0) = 1$.
3. As the solution of the IVP: $dy/dx = y$, $y(0) = 1$.

These definitions are not completely equivalent: the last one requires differentiability while the first two do not. Nor are they as convenient as (5.2).

Figure 5.1 was inspired by one of the many masterful illustrations in [2]. Although we will not use complex analysis, it may be helpful in understanding the Fourier and Laplace transforms discussed in Chap. 10. Even more helpful will be Nahin's gem of a book [1], particularly its Chaps. 4–6.

5.6 Exercises

1. Show that a complex number $x + y i$ can be written in polar form $r e^{i\theta}$ and use that to define $\ln(z)$ (complex logarithm is multi-valued and you will need to define its principal branch). Find $\ln(-1)$ symbolically and validate it in MATLAB.
2. Derive an analog of Euler's formula (5.5) for $\sin(z) = \sin(x + y i)$ and use it to produce a conformal plot of $\sin(z)$.
3. Use Euler's formula (5.5) to derive identities for cosine and sine of the triple angle. *Hint:* Compare $e^{3i\theta}$ to $(e^{i\theta})^3$.

4. Find all complex solutions of the equation $e^z = 1$. This exercise will be useful in Sect. 9.5 where we solve the heat equation with Neumann boundary conditions.
5. Does the property $e^x e^y = e^{x+y}$ hold for matrix exponentials? Either give a proof that it does, or provide a counterexample.
6. Exponentiate the following matrices using the definition (5.10):

$$\begin{bmatrix} a & 0 \\ 0 & b \end{bmatrix}, \begin{bmatrix} a & 0 \\ b & 0 \end{bmatrix}, \begin{bmatrix} a & 0 \\ b & 1 \end{bmatrix}.$$

Which of the three matrices is the easiest to exponentiate?

7. In Sect. 4.4 we showed that mutarotation of glucose can be modeled with the following ODE (Eq. (4.13)):

$$\frac{d}{dt}\begin{bmatrix} x \\ y \end{bmatrix} = \begin{bmatrix} -k_1 & k_2 \\ k_1 & -k_2 \end{bmatrix}\begin{bmatrix} x \\ y \end{bmatrix}.$$

As follows from the discussion in Sect. 5.4, the solution can be written in matrix form as

$$\begin{bmatrix} x \\ y \end{bmatrix} = \exp\left(t \begin{bmatrix} -k_1 & k_2 \\ k_1 & -k_2 \end{bmatrix}\right) \begin{bmatrix} x_0 \\ y_0 \end{bmatrix}$$

The following code computes the solution using expm and adds a bit of noise.

```
k1 = 1.3; k2 = .7; A = [-k1 k2; k1 -k2];
t = linspace(0,5); u0 = [1; 0];
u = zeros(length(t),2);
u(1,:) = u0;
for n = 2:length(t)
    u(n,:) = expm(t(n)*A)*u0;
end
u = u + .01*randn(size(u));
```

Using the output of the code as data, estimate the rate constants k_1 and k_2.

References

1. P.J. Nahin, *Dr. Euler's Fabulous Formula: Cures Many Mathematical Ills*. (Princeton University Press, 2017)
2. T. Needham, *Visual Complex Analysis* (Clarendon Press, 1997)

ODE and Linear Algebra   6

Except for the logistic equation in Chap. 1, all ODE models in the preceding chapters are linear. We prioritize linear ODE for two reasons: firstly, as shown in Chaps. 3 and 4, many ODE models are either linearized or are linear to begin with; secondly, and perhaps more importantly, linear ODE theory is a necessary prerequisite for nonlinear ODE theory.

In this chapter we rigorously define linearity and show that it is a constructive property: if an ODE is linear then its general solution has certain structure. That structure is codified in the structure theorem (Sect. 6.6) which, together with the principle of superposition (Sect. 6.8), provides a blueprint for constructing the general solution of a linear ODE. In contrast, there is no blueprint for solving a nonlinear ODE: if separation of variables and other Calculus tricks do not work, the only recourse is numerical approximation.

Linearity is not specific to differential equations: it is a property of transformations between vector spaces. In order to define linearity of ODE (Sect. 6.5), we must first define vector spaces, linear transformations, and some auxiliary algebraic concepts, which we do in the next four sections.

6.1 Number Fields

Expounding upon number fields in the middle of a book about differential equations may seem out of place, but it is not: without number fields there are no vector spaces, without vector spaces there are no linear transformations, and without linear transformations there are no linear ODE. Defining number fields is also useful for dispelling lingering doubts about the validity of the complex number system. These doubts, rooted in identification of real numbers with symbols for measuring and counting, undermine the understanding of complex exponentials and the best way to put them to rest is with a formal definition.

A number field is a set (with elements a, b, c, etc.) endowed with operations of addition $+$ and multiplication $\times$. These operations satisfy three groups of axioms: the axioms of addition, the axioms of multiplication, and a distributive law.

Number addition is associative and commutative; there is an additive identity, called "zero"; every number has an additive inverse:

(NF1) Associativity of addition: $(a + b) + c = a + (b + c)$.
(NF2) Commutativity of addition: $a + b = b + a$.
(NF3) There is a number 0 such that $a + 0 = a$ for all a.
(NF4) For every a there exists b such that $a + b = 0$.

Subtraction in a number field is defined in terms of addition: the difference $a - b$ is the number c such that $a = b + c$; the additive inverse of a is customarily denoted $-a$.

The axioms of multiplication parallel those of addition: number multiplication is associative and commutative; there is a multiplicative identity, called "one"; every number, except for zero, has a multiplicative inverse:

(NF5) Associativity of multiplication: $(a \times b) \times c = a \times (b \times c)$.
(NF6) Commutativity of multiplication: $a \times b = b \times a$.
(NF7) There is a number 1 such that $a \times 1 = a$ for all a.
(NF8) For every $a \neq 0$ there exists b such that $a \times b = 1$.

Just as subtraction is defined in terms of addition, division is defined in terms of multiplication: the ratio a/b is the number c such that $a = b \times c$. Since $0 \times c = 0$ for all c, division by 0 is not allowed.

To be compatible, the operations $+$ and $\times$ must satisfy the distributive law:

(NF9) Multiplication is distributive over addition: $a \times (b + c) = (a \times b) + (a \times c)$.

Any set with $+$ and $\times$ satisfying axioms (NF1)–(NF9) is a number field. Numbers are elements of number fields: they are defined by arithmetic properties of addition and multiplication rather than attributes or practical use.

Not every set of numbers is a number field. For instance, the set of all integers $\mathbb{Z}$ is not a field because the quotient of two integers is not necessarily an integer (the set $\mathbb{Z}$ with the operations $+$ and $\times$ has the structure of a *ring*). However, if all quotients of integers are adjoined to $\mathbb{Z}$ then it becomes the field of rational numbers $\mathbb{Q}$.

In Calculus $\mathbb{Q}$ is enlarged by adjoining the limits of all convergent sequences of rational numbers: this gives the field of real numbers $\mathbb{R}$. Adjoining $i = \sqrt{-1}$ to $\mathbb{R}$ (and all numbers that can be formed from i and real numbers through arithmetic operations) gives the field of complex numbers $\mathbb{C}$.

Unfortunately, complex numbers clash with our innate number sense. Rules like "squares are positive" are so ingrained during primary education that years later our subconsciousness stubbornly reminds us that i^2 cannot possibly equal -1. Yet it does and there is no contradiction: complex numbers contain real numbers as a subfield and therefore it is the real numbers that must comport with statements about complex numbers rather than the other way around. Complex numbers may appear strange but they are not controversial: since complex addition and multiplication comply with (NF1)–(NF9), $\mathbb{C}$ is a number field and its elements are bona fide numbers.

6.2 Vector Spaces

In early Physics and Calculus, vectors are introduced as arrows representing "quantities with direction and magnitude." At a certain point arrows give way to linear combinations of $\{\mathbf{i}, \mathbf{j}, \mathbf{k}\}$ and columns (or, more often, rows) of real numbers. Yet, the early association

$$\text{vector} \leftrightarrow \text{magnitude} + \text{direction}$$

persists and gets in the way of understanding the true nature of vectors which resides not in vector attributes but in vector operations.

Just as numbers are defined collectively, as number fields, vectors (denoted $\mathbf{x}, \mathbf{y}, \mathbf{z}$, etc.) are defined collectively as vector spaces. A vector space is a set with vector addition $\oplus$ and scalar multiplication $\otimes$. These operations satisfy a set of axioms which, just like the number field axioms, may be divided into three groups: (i) axioms specific to vector addition; (ii) axioms specific to scalar multiplication; and (iii) axioms describing interactions between vector addition and scalar multiplication.

Vector addition is a binary operation that takes two vectors as inputs and produces a vector as an output. It is required to be associative and commutative; there is a "zero vector" serving as the additive identity; every vector has an additive inverse:

(VS1) Associativity of vector addition: $(\mathbf{x} \oplus \mathbf{y}) \oplus \mathbf{z} = \mathbf{x} \oplus (\mathbf{y} \oplus \mathbf{z})$.
(VS2) Commutativity of vector addition: $\mathbf{x} \oplus \mathbf{y} = \mathbf{y} \oplus \mathbf{x}$.
(VS3) There is a vector $\mathbf{0}$ such that $\mathbf{x} \oplus \mathbf{0} = \mathbf{x}$.
(VS4) For every $\mathbf{x}$ there exists $\mathbf{y}$ such that $\mathbf{x} \oplus \mathbf{y} = \mathbf{0}$.

Scalar multiplication (scaling) is multiplication of vectors by numbers (scalars) from the underlying number field: vector spaces are defined "over" number fields. As far as scalar multiplication is concerned, we need to spell out what happens when we scale a vector by a product of scalars; also, we must stipulate that scaling by 1 does not alter the vector:

(VS5) Scaling by a product of scalars is repeated scaling: $(a \times b) \otimes \mathbf{x} = a \otimes (b \otimes \mathbf{x})$.
(VS6) Scaling by 1 does not alter the vector: $1 \otimes \mathbf{x} = \mathbf{x}$.

Finally, we need distributive rules for scaling a sum of vectors by a scalar and for scaling a vector by a sum of scalars:

(VS7) Scaling a vector sum: $a \otimes (\mathbf{x} \oplus \mathbf{y}) = (a \otimes \mathbf{x}) \oplus (a \otimes \mathbf{y})$.
(VS8) Scaling by a sum of scalars: $(a+b) \otimes \mathbf{x} = (a \otimes \mathbf{x}) \oplus (b \otimes \mathbf{x})$.

Any set with addition and scalar multiplication satisfying (VS1)–(VS8) is a vector space (over the prescribed number field). Vectors are elements of vector spaces, which is to say, objects that can be added and scaled in accordance with the vector space axioms. Notice that there is nothing in (VS1)–(VS8) that suggests magnitudes or directions. In fact, unless a vector space is given a *norm* or an *inner product* its vectors are completely devoid of geometric characteristics.

The reader must have noticed close similarity between the axioms of vector addition (VS1)–(VS4) and those of number addition (NF1)–(NF4). Both sets of axioms define an algebraic structure called an *abelian group*: vector spaces are abelian groups with respect to $\oplus$ while number fields are abelian groups with respect to $+$. Since $\oplus$ and $+$ are really one and the same operation but applied to different objects, we will use $+$ for both number and vector addition reserving $\oplus$ for rare situations where a distinction must be made, as in axioms (VS1)–(VS8). Following common mathematical practice, we will also omit $\otimes$ and $\times$ in formulas, as we have been doing, unless there is a compelling reason to do otherwise.

What follows are several examples of vector spaces over the reals. The last example illustrates the connection to linear ODE.

We start with the most important vector space, $\mathbb{R}^n$, which should be familiar from Calculus. By definition, the vectors in $\mathbb{R}^n$ are n-tuples of real numbers which are added and scaled componentwise. We arrange these n-tuples as columns because of the row-by-column rule of matrix multiplication; to save space, we may sometimes typeset a column as a transposed row.

In Calculus, $\mathbb{R}^n$ is equipped with the dot product—a special case of an inner product. Unlike vector addition and scalar multiplication, the dot product is not a native vector space operation—it is introduced separately from the vector space axioms. The purpose of the dot product is to define lengths and angles via

$$\|\mathbf{x}\| = \sqrt{\mathbf{x} \cdot \mathbf{x}}, \quad \cos(\theta) = \frac{\mathbf{x} \cdot \mathbf{y}}{\|\mathbf{x}\| \, \|\mathbf{y}\|},$$

thereby turning $\mathbb{R}^n$ into *Euclidean space*.

As the next example of a real vector space, consider the set of polynomials of degree n with real coefficients: $P_n = \left\{ \sum_{k=0}^n a_k x^k \mid a_k \in \mathbb{R} \right\}$. The addition and scaling of polynomials

$$\sum_{k=0}^n a_k x^k + \sum_{k=0}^n b_k x^k = \sum_{k=0}^n (a_k + b_k) x^k, \quad c \sum_{k=0}^n a_k x^k = \sum_{k=0}^n (c \, a_k) x^k$$

6.2 Vector Spaces

is very similar to that of vectors in $\mathbb{R}^{n+1}$, so much so that we can identify

$$\sum_{k=0}^{n} a_k x^k \leftrightarrow \begin{bmatrix} a_n & a_{n-1} & \cdots & a_0 \end{bmatrix}^T.$$

The technical term for such identifications is *isomorphism*. MATLAB uses this particular isomorphism between P_n and $\mathbb{R}^{n+1}$ in commands like `polyval` which expect a polynomial to be input as a vector of coefficients listed in descending order.

We will elaborate on isomorphisms of vector spaces in Sect. 6.4, after defining linear transformations. In the meantime, as another illustration of the concept, consider the set of m-by-n matrices with real entries, $M_{m \times n}(\mathbb{R})$. Since matrices are added and scaled componentwise, $M_{m \times n}(\mathbb{R})$ is isomorphic to $\mathbb{R}^{m \times n}$. In MATLAB this isomorphism is implemented as the colon operator: if A is an m-by-n matrix then A(:) is the column vector obtained by stacking the columns of A.

Certain linear algebra definitions and theorems reference the *trivial vector space*. This is the simplest vector space consisting of just the zero vector. Often, the trivial vector space $\{0\}$ is the designation for the *subspace* of a given vector space consisting of its zero vector; a subspace of a vector space is a subset that is a vector space in its own right.

We conclude with a vector space directly related to an ODE. Let V be the general solution—the set of all solutions of the natural growth equation $dP/dt = a P$. In Sect. 1.2 we showed that $V = \{C e^{at} \mid C \in \mathbb{R}\}$. This set is closed under addition and scaling by real numbers and, therefore, is a vector space (isomorphic to $\mathbb{R}$). Yet, even if we did not know that V consists of scalar multiples of e^{at}, we could still show that it is a vector space as follows.

Let P_1 and P_2 be any two solutions of the natural growth equation: this means that $dP_1/dt = a P_1$ and $dP_2/dt = a P_2$. Set $P = P_1 + P_2$. Then

$$\frac{dP}{dt} = \frac{dP_1}{dt} + \frac{dP_2}{dt} = a P_1 + a P_2 = a (P_1 + P_2) = a P,$$

so the sum of solutions is another solution. Likewise, if P is a solution of $dP/dt = a P$ and c is a real number then

$$\frac{d}{dt}(c P) = c \frac{dP}{dt} = c a P = a (c P),$$

so scaling a solution gives another solution. Since the elements of V can be added and scaled in compliance with vector space axioms (VS1)—(VS8), V is a vector space.

Due to the early association of vectors with arrows, the notion that polynomials, exponentials, and other functions are vectors is often met with incredulity. Functions do not look like arrows. Nor do they have visible magnitudes and directions, like arrows. Nevertheless, functions add and scale like vectors and therefore are vectors.

6.3 Bases

Computations in a vector space usually require a *basis*. The definition of a basis involves the terms *linear independence* and *span* which must be defined first.

A set of vectors $\{\mathbf{v}_k\}_{k=1}^n$ is *linearly independent* if $\sum_{k=1}^n c_k \mathbf{v}_k = \mathbf{0}$ implies that all c_k's are zero: the only vanishing linear combination of linearly independent vectors is the *trivial* one.

The vectors $\{\mathbf{i}, \mathbf{j}, \mathbf{k}\}$ exemplify linearly independent vectors in $\mathbb{R}^n$ ($n \geq 3$). Meanwhile, if $\mathbf{l} = \mathbf{i} + \mathbf{j} + \mathbf{k}$ then $\{\mathbf{i}, \mathbf{j}, \mathbf{k}, \mathbf{l}\}$ are linearly dependent because $\mathbf{l} - \mathbf{i} - \mathbf{j} - \mathbf{k} = \mathbf{0}$ is a nontrivial vanishing linear combination.

As an example closer to ODE, consider the vector space of complex-valued functions (of time) defined over complex numbers. The sets of real harmonics $\{\cos(t), \sin(t)\}$ and complex harmonics $\{e^{it}, e^{-it}\}$ are linearly independent, but $\{\cos(t), \sin(t), e^{it}\}$ is not because $e^{it} - \cos(t) - i\sin(t) = 0$ in accordance with Euler's formula (5.5).

The span of a collection of vectors (which may or may not be linearly independent) is the set of all linear combinations:

$$\text{span}\left(\{\mathbf{v}_k\}_{k=1}^n\right) = \left\{\sum_{k=1}^n c_k \mathbf{v}_k\right\}.$$

Span is always a subspace: in fact, $\text{span}\left(\{\mathbf{v}_k\}_{k=1}^n\right)$ is the smallest subspace containing $\{\mathbf{v}_k\}_{k=1}^n$. For instance, the span of $\{\mathbf{i}, \mathbf{j}, \mathbf{k}\}$ is the three-dimensional subspace of $\mathbb{R}^n$ ($n \geq 3$) consisting of vectors of the form $a\,\mathbf{i} + b\,\mathbf{j} + c\,\mathbf{k}$; if taken in $\mathbb{R}^3$ it is all of $\mathbb{R}^3$. The span of $\{\mathbf{i}, \mathbf{j}, \mathbf{k}, \mathbf{l}\}$ where $\mathbf{l} = \mathbf{i} + \mathbf{j} + \mathbf{k}$ is exactly the same as the span of $\{\mathbf{i}, \mathbf{j}, \mathbf{k}\}$ because

$$c_1 \mathbf{i} + c_2 \mathbf{j} + c_3 \mathbf{k} + c_4 \mathbf{l} = (c_1 + 1)\mathbf{i} + (c_2 + 1)\mathbf{j} + (c_3 + 1)\mathbf{k} \in \text{span}\left(\{\mathbf{i}, \mathbf{j}, \mathbf{k}\}\right).$$

Likewise, the spans of $\{\cos(t), \sin(t)\}$ and $\{e^{it}, e^{-it}\}$ are the same if the coefficients of linear combinations are complex numbers.

Now, a basis of a vector space V is a set of linearly independent vectors $\{\mathbf{e}_k\}_{k=1}^n$ that span V. The spanning requirement means that every $\mathbf{v} \in V$ can be written as a linear combination

$$\mathbf{v} = \sum_{k=1}^n c_k \mathbf{e}_k. \tag{6.1}$$

Meanwhile, the linear independence requirement ensures that (6.1) is unique. Indeed, if in addition to (6.1) we can also write $\mathbf{v} = \sum_{k=1}^n c'_k \mathbf{e}_k$ then we can form a vanishing linear combination $\sum_{k=1}^n (c'_k - c_k) \mathbf{e}_k = \mathbf{0}$. Yet, since $\{\mathbf{e}_k\}_{k=1}^n$ are linearly independent, their only vanishing combination is trivial, so $c'_k = c_k$ for all $k = 1, \ldots, n$.

Writing vectors as linear combinations of basis vectors is called *basis representation* or *basis expansion*. The coefficients c_k in (6.1) are the *components of* $\mathbf{v}$ *with respect to the basis* $\{\mathbf{e}_k\}_{k=1}^n$. If the basis is changed, the components also change. For instance, the components of $\mathbf{l} = \mathbf{i}+\mathbf{j}+\mathbf{k}$ with respect to the standard basis $\{\mathbf{i},\mathbf{j},\mathbf{k}\}$ of $\mathbb{R}^3$ are $\begin{bmatrix} 1 & 1 & 1 \end{bmatrix}^T$ but with respect to the basis $\{\mathbf{i}-\mathbf{j},\mathbf{i}-\mathbf{k},\mathbf{i}+2\mathbf{j}+\mathbf{k}\}$ they are $\begin{bmatrix} 1/2 & -1/4 & 3/4 \end{bmatrix}^T$. The only vector whose components do not depend on the choice of basis is the zero vector: all components of $\mathbf{0}$ are always zero.

With the exception of the trivial vector space, every vector space has multiple bases. The number of basis vectors, however, is always the same and is, by definition, the *dimension* of the vector space. Thus $\mathbb{R}^3$ is three-dimensional because its standard basis $\{\mathbf{i},\mathbf{j},\mathbf{k}\}$ has three vectors, as does every other of its infinitely many bases. The space of solutions of the natural growth equation $V = \{C\,e^{at} \mid C \in \mathbb{R}\}$ has basis $\{e^{at}\}$ and is therefore one-dimensional; every other basis of V, such as $\{2\,e^{at}\}$ or $\{-3\,e^{at}\}$, also consists of a single element.

Including the zero vector in a basis is not allowed because that violates the requirement of linear independence of basis vectors. This means that the basis of the trivial vector space $\{\mathbf{0}\}$ must be the empty set $\{\ \}$ and the dimension of the trivial vector space is therefore zero. At the opposite extreme, the dimension of a vector space may be infinite—some functional inner product spaces in later chapters are infinite-dimensional.

Once a basis in a real vector space is fixed, the vectors in that space may be written as columns of components. This turns the space into a copy of $\mathbb{R}^n$—we saw two examples of that in Sect. 6.2 where we discussed the space of polynomials P_n and the space of matrices $M_{m \times n}(\mathbb{R})$. As we explain in the next section, using bases also turns linear transformations into matrices. In fact, this is what matrices are in linear algebra—representations of linear transformations.

6.4 Linear Transformations

Let $T : V \to W$ be a transformation from a vector space V into a vector space W. T is said to be linear if it can be distributed over linear combinations:

$$T\left(\sum_{k=1}^{n} c_k\,\mathbf{v}_k\right) = \sum_{k=1}^{n} c_k\,T(\mathbf{v}_k). \tag{6.2}$$

In practice, we will test transformations for linearity by applying them to linear combinations of just two vectors: if $T(c_1\,\mathbf{v}_1 + c_2\,\mathbf{v}_2) = c_1\,T(\mathbf{v}_1) + c_2\,T(\mathbf{v}_2)$ holds for arbitrary $\mathbf{v}_1$ and $\mathbf{v}_2$ in V then T can be distributed over arbitrary finite linear combinations in V, as in definition (6.2).

Suppose that $T : V \to W$ is a linear transformation. Fix a basis $\{\mathbf{v}_j\}_{j=1}^{n}$ in V and, independently, a basis $\{\mathbf{w}_i\}_{i=1}^{m}$ in W. Then

$$T(\mathbf{v}) = T\underbrace{\left(\sum_{j=1}^{n} c_j\, \mathbf{v}_j\right)}_{\text{basis expansion of } \mathbf{v}} = \underbrace{\sum_{j=1}^{n} c_j\, T(\mathbf{v}_j)}_{\text{linearity of } T} = \sum_{j=1}^{n} c_j \underbrace{\sum_{i=1}^{m} a_{ij}\, \mathbf{w}_i}_{\text{basis expansion of } T(\mathbf{v}_j)}$$

$$= \sum_{i=1}^{m} \left(\sum_{j=1}^{n} a_{ij}\, c_j\right) \mathbf{w}_i,$$

which shows that T transforms a V-vector with components c_j into a W-vector with components $\sum_{j=1}^{n} a_{ij}\, c_j$. If we write vectors as columns of components then we can compute $T(\mathbf{v})$ as the matrix-vector product $A\,\mathbf{c}$ where $A = [a_{ij}]$ is the *matrix representation* of T. Of course, this is not a coincidence: we write vectors as columns and define matrix multiplication using the row-by-column rule in order to be able to do just that. Furthermore, as we are about to demonstrate, the row-by-column rule is consistent with composition of linear transformations.

As an illustration of matrix representation, let us consider the action of the derivative d/dx on spaces of polynomials P_n with bases of monomials arranged in descending order $\{x^n, x^{n-1}, \ldots, 1\}$—as we mentioned in Sect. 6.2, these bases are used in MATLAB.

Differentiating a cubic

$$\frac{d}{dx} : a_3\, x^3 + a_2\, x^2 + a_1\, x + a_0 \mapsto 3\, a_3\, x^2 + 2\, a_2\, x + a_1$$

is equivalent to multiplying a vector in $\mathbb{R}^4$ by a 3-by-4 matrix

$$\frac{d}{dx} : \begin{bmatrix} a_3 \\ a_2 \\ a_1 \\ a_0 \end{bmatrix} \mapsto \begin{bmatrix} 3\, a_3 \\ 2\, a_2 \\ a_1 \end{bmatrix} = \begin{bmatrix} 3 & 0 & 0 & 0 \\ 0 & 2 & 0 & 0 \\ 0 & 0 & 1 & 0 \end{bmatrix} \begin{bmatrix} a_3 \\ a_2 \\ a_1 \\ a_0 \end{bmatrix}.$$

Similarly, differentiating a quadratic

$$\frac{d}{dx} : a_2\, x^2 + a_1\, x + a_0 \mapsto 2\, a_2\, x + a_1$$

is equivalent to multiplying a vector in $\mathbb{R}^3$ by a 2-by-3 matrix

$$\frac{d}{dx} : \begin{bmatrix} a_2 \\ a_1 \\ a_0 \end{bmatrix} \mapsto \begin{bmatrix} 2\, a_2 \\ a_1 \end{bmatrix} = \begin{bmatrix} 2 & 0 & 0 \\ 0 & 1 & 0 \end{bmatrix} \begin{bmatrix} a_2 \\ a_1 \\ a_0 \end{bmatrix}.$$

6.4 Linear Transformations

Thus the two matrices $\begin{bmatrix} 3 & 0 & 0 & 0 \\ 0 & 2 & 0 & 0 \\ 0 & 0 & 1 & 0 \end{bmatrix}$ and $\begin{bmatrix} 2 & 0 & 0 \\ 0 & 1 & 0 \end{bmatrix}$ represent linear transformations d/dx : $P_3 \to P_2$ and $d/dx : P_2 \to P_1$, respectively (despite being defined by the same Calculus derivative, these transformations are distinct because they involve different vector spaces).

The matrix product

$$\begin{bmatrix} 2 & 0 & 0 \\ 0 & 1 & 0 \end{bmatrix} \begin{bmatrix} 3 & 0 & 0 & 0 \\ 0 & 2 & 0 & 0 \\ 0 & 0 & 1 & 0 \end{bmatrix} = \begin{bmatrix} 6 & 0 & 0 & 0 \\ 0 & 2 & 0 & 0 \end{bmatrix}$$

gives the matrix representation of $d/dx \circ d/dx = d^2/dx^2 : P_3 \to P_1$: the row-by-column rule of matrix multiplication corresponds to composition of linear transformations.

Matrix representation of a linear transformation generally depends on the choice of two bases and different choices give different matrices. For instance, if we order the basis monomials in increasing order $\{1, x, \ldots, x^n\}$ then the matrix representations of d/dx : $P_3 \to P_2$ and $d/dx : P_2 \to P_1$ become $\begin{bmatrix} 1 & 0 & 0 & 0 \\ 0 & 2 & 0 & 0 \\ 0 & 0 & 3 & 0 \end{bmatrix}$ and $\begin{bmatrix} 1 & 0 & 0 \\ 0 & 2 & 0 \end{bmatrix}$

Matrix representation of a *linear operator*—a linear transformation from a vector space into itself—can be constructed using a single basis. For instance, if we identify P_2 with a subspace of P_3 via

$$a_2 x^2 + a_1 x + a_0 = 0 x^3 + a_2 x^2 + a_1 x + a_0$$

then the matrix of $d/dx : P_3 \to P_3$ with respect to the basis of descending monomials $\{x^3, x^2, x, 1\}$ is $\begin{bmatrix} 0 & 0 & 0 & 0 \\ 3 & 0 & 0 & 0 \\ 0 & 2 & 0 & 0 \\ 0 & 0 & 1 & 0 \end{bmatrix}$. One of the central questions in linear algebra is how to choose bases to make the matrix representation of a linear transformation or an operator as simple as possible. As we will explain in Chap 7, for a linear operator the most natural basis is the set of its *eigenvectors*.

While discussing the trivial vector space $\{0\}$ in Sect. 6.2 we introduced the term "subspace" for a subset of a vector space that is closed under vector space operations and is therefore a vector space itself. Associated with every linear transformation are two fundamental subspaces called the *null space* and the *range*. The null space is particularly important because it features in Theorem 6.1 of Sect. 6.6—the main theorem of the chapter.

The null space of $T : V \to W$ is the set of all vectors in V that are mapped to zero in W: $N(T) = \{\mathbf{v} \in V \mid T(\mathbf{v}) = \mathbf{0} \in W\}$. If $\{\mathbf{v}_k\}_{k=1}^n$ is a collection of vectors in $N(T)$ then, by definition, $T(\mathbf{v}_k) = \mathbf{0}$ for all $k = 1, \ldots, n$ and, by linearity of T, $T\left(\sum_{k=1}^n c_k \mathbf{v}_k\right) = \sum_{k=1}^n c_k T(\mathbf{v}_k) = \mathbf{0}$. This shows that the set $N(T)$ is closed under linear combinations in V and therefore is a subspace of V.

It may happen that $N(T)$ consists only of the zero vector in which case it is said to be *trivial*. Linear transformations with trivial null spaces are *injective* or *one-to-one*: they map different inputs into different outputs. Indeed, suppose that $T(\mathbf{v}_1) = T(\mathbf{v}_2)$. By linearity, $T(\mathbf{v}_1) - T(\mathbf{v}_2) = T(\mathbf{v}_1 - \mathbf{v}_2)$ and since $T(\mathbf{v}_1) - T(\mathbf{v}_2) = \mathbf{0}$ this means that $\mathbf{v}_1 - \mathbf{v}_2 \in N(T)$. Now if $N(T) = \{\mathbf{0}\}$ then $\mathbf{v}_1 - \mathbf{v}_2 = \mathbf{0}$. Consequently, $T(\mathbf{v}_1) = T(\mathbf{v}_2)$ implies $\mathbf{v}_1 = \mathbf{v}_2$ thereby proving injectivity.

The range of $T : V \to W$ is the image of V under T: $R(T) = \{T(\mathbf{v}) \in W \mid \mathbf{v} \in V\}$. We leave it as an exercise to show that $R(T)$ is a vector subspace of W. If $R(T) = W$ then the transformation T is said to be *surjective* or *onto*.

In Sect. 6.2 we described isomorphisms between the spaces of polynomials P_n and $\mathbb{R}^{n+1}$ and between the spaces of real-valued matrices $M_{m \times n}(\mathbb{R})$ and $\mathbb{R}^{m \times n}$. We can now rigorously define isomorphism between vector spaces V and W as a linear transformation $T : V \to W$ that is both injective (one-to-one) and surjective (onto). Isomorphic vector spaces differ in notation only: from the point of view of linear algebra, they are one and the same. We should also add that the idea of isomorphism is not restricted to vector spaces. Quite generally, an isomorphism between two algebraic objects is an injective and surjective (bijective) map that preserves algebraic operations. For instance, if F_1 and F_2 are two number fields, then $\phi : F_1 \to F_2$ is an isomorphism if, in addition to being injective and surjective, it preserves addition and multiplication:

$$\phi(a+b) = \phi(a) + \phi(b), \quad \phi(a\,b) = \phi(a)\,\phi(b).$$

An example of isomorphism of number fields is given in the Comments and Bibliography section.

6.5 Linear Equations

Any system of equations, differential or not, can be cast into the form

$$T(x) = y \tag{6.3}$$

where T is a transformation taking the unknown x into some known quantity y. If T is a linear transformation between vector spaces X and Y then Eq. (6.3) is linear. If, moreover, $y = 0$ then (6.3) is *linear homogeneous* (the emphasis is on the third syllable: *homogéneous*); otherwise it is *linear nonhomogeneous*.

We will first look at examples of linear differential equations, since they are our main concern.

In Sect. 1.2 we stated that the natural growth equation $dP/dt = a\,P$ may be classified as linear homogeneous because the derivative dP/dt is directly proportional to P. That is how linearity is recognized, but the true reason why the natural growth equation is linear homogeneous is that it can be written as $T(P) = 0$ with $T : P \mapsto dP/dt - a\,P$ a linear

6.5 Linear Equations

transformation from the space of continuously differentiable functions $C^1(\mathbb{R})$ to the space of continuous functions $C(\mathbb{R})$. To check the linearity of T, apply it to a linear combination of two arbitrary functions:

$$T(c_1 P_1 + c_2 P_2) = \frac{d}{dt}(c_1 P_1 + c_2 P_2) - a(c_1 P_1 + c_2 P_2)$$

$$= c_1 \left(\frac{dP_1}{dt} - a P_1\right) + c_2 \left(\frac{dP_2}{dt} - a P_2\right) = c_1 T(P_1) + c_2 T(P_2).$$

Since T can be distributed over linear combinations, it is linear and so is the natural growth equation defined by it.

The general form of a scalar linear ODE of order n is

$$a_n(t) x^{(n)} + a_{n-1}(t) x^{(n-1)} + a_{n-2}(t) x^{(n-2)} + \cdots + a_0(t) x = f(t). \quad (6.4)$$

Linear[ized] ODE in Chap. 3 are special cases of (6.4) with $n = 1$

Equation (6.4) can be further generalized by replacing the scalar unknown $x(t)$ with a vector unknown $\mathbf{x}(t)$ and the scalar coefficients $a_k(t)$ with matrices $A_k(t)$:

$$A_n(t) \mathbf{x}^{(n)} + A_{n-1}(t) \mathbf{x}^{(n-1)} + A_{n-2}(t) \mathbf{x}^{(n-2)} + \cdots + A_0(t) \mathbf{x} = \mathbf{f}(t). \quad (6.5)$$

Equation (5.8) from Sect. 5.4 is an important special case of Eq. (6.5).

Henceforth we assume that $a_n(t) \neq 0$ in (6.4) and the inverse matrix $A_n^{-1}(t)$ exists in (6.5) so that we can divide by the leading coefficients. With these assumptions we can recast (6.4) and (6.5) as

$$\frac{d\mathbf{u}}{dt} = M(t)\mathbf{u} + \mathbf{g}(t),$$

where $\mathbf{u}$ is a vector built from the unknown and its derivatives of orders up to $n - 1$. For instance, consider ODE (6.4) with $n = 3$. Set $\mathbf{u} = \begin{bmatrix} x & x^{(1)} & x^{(2)} \end{bmatrix}^T$. Then the derivative of $\mathbf{u}$ is

$$\frac{d\mathbf{u}}{dt} = \begin{bmatrix} x^{(1)} \\ x^{(2)} \\ x^{(3)} \end{bmatrix} = \underbrace{\begin{bmatrix} x^{(1)} \\ x^{(2)} \\ \left(f(t) - a_0(t) x - a_1(t) x^{(1)} - a_2(t) x^{(2)}\right)/a_3(t) \end{bmatrix}}_{\text{solve } a_3(t) x^{(3)} + a_2(t) x^{(2)} + a_1(t) x^{(1)} + a_0(t) x = f(t) \text{ for } x^{(3)}}$$

$$= \begin{bmatrix} 0 & 1 & 0 \\ 0 & 0 & 1 \\ -\frac{a_0(t)}{a_3(t)} & -\frac{a_1(t)}{a_3(t)} & -\frac{a_2(t)}{a_3(t)} \end{bmatrix} \begin{bmatrix} x \\ x^{(1)} \\ x^{(2)} \end{bmatrix} + \begin{bmatrix} 0 \\ 0 \\ \frac{f(t)}{a_3(t)} \end{bmatrix} = M(t)\mathbf{u} + \mathbf{g}(t).$$

Similarly, in order to convert to first order Eq. (6.5) with $n = 3$, stack the vectors $\mathbf{x}$, $\mathbf{x}^{(1)}$, and $\mathbf{x}^{(2)}$ into a single vector $\mathbf{u}$ and compute the derivative:

$$\frac{d\mathbf{u}}{dt} = \begin{bmatrix} \mathbf{x}^{(1)} \\ \mathbf{x}^{(2)} \\ \mathbf{x}^{(3)} \end{bmatrix} = \begin{bmatrix} \mathbf{x}^{(1)} \\ \mathbf{x}^{(2)} \\ \underbrace{A_3^{-1}(t) \left(\mathbf{f} - A_0(t)\,\mathbf{x} - A_1(t)\,\mathbf{x}^{(1)} - A_2(t)\,\mathbf{x}^{(2)} \right)}_{\text{solve } A_3(t)\,\mathbf{x}^{(3)} + A_2(t)\,\mathbf{x}^{(2)} + A_1(t)\,\mathbf{x}^{(1)} + A_0(t)\,\mathbf{x} = \mathbf{f}(t) \text{ for } \mathbf{x}^{(3)}} \end{bmatrix}$$

$$= \begin{bmatrix} O & I & O \\ O & O & I \\ -A_3^{-1}(t)\,A_0(t) & -A_3^{-1}(t)\,A_1(t) & -A_3^{-1}(t)\,A_2(t) \end{bmatrix} \begin{bmatrix} \mathbf{x} \\ \mathbf{x}^{(1)} \\ \mathbf{x}^{(2)} \end{bmatrix}$$

$$+ \begin{bmatrix} \mathbf{0} \\ \mathbf{0} \\ A_3^{-1}(t)\,\mathbf{f} \end{bmatrix} = M(t)\,\mathbf{u} + \mathbf{g}(t).$$

In the preceding equation the zero vector $\mathbf{0}$ in $\mathbf{g}(t)$ has the same size as $\mathbf{u}$ while O and I in $M(t)$ have the same dimensions as A_k's: O is the matrix of zeros and I is the *identity matrix*—the diagonal matrix with 1's on the main diagonal.

We conclude with two examples of linear equations which, despite not being differential, play an important role in ODE theory.

Let A be an m-by-n matrix with real or complex entries. The algebraic equation $A\,\mathbf{x} = \mathbf{b}$ is an archetype of a linear equation: when ODE (and other kinds of linear equations) are discretized, they assume this form.

To motivate separation of variables in Sect. 1.3, we showed that indefinite integration of both sides of the natural growth equation leads to a dead end. Integrating with limits, however, leads to an *integral equation*

$$P(t) = P_0 + a \int_0^t P(s)\,ds \tag{6.6}$$

whose solution is the same as that of the IVP

$$\frac{dP}{dt} = a\,P, \quad P(0) = P_0.$$

In the same way n-fold definite integration converts ODE (6.4) into

$$x(t) = g(t) + \int_0^t K(t, s)\,x(s)\,ds, \tag{6.7}$$

which is the general form of a *Volterra integral equation of the second kind*. While it is usually simpler to solve an ODE (analytically or numerically), proving that it has a solution and that the solution is unique is usually simpler for the equivalent integral equation.

6.6 Structure Theorem

Let $T : X \to Y$ be a linear transformation. As follows from the definition of the null space given in Sect. 6.4, the general solution of the linear homogeneous equation $T(\mathbf{x}) = \mathbf{0}$ is the null space $N(T)$ which is a vector subspace of X. The general solution of the nonhomogeneous equation $T(\mathbf{x}) = \mathbf{y}$ is not a vector space but it is not far removed.

Theorem 6.1 (Structure theorem) *The general solution of the linear nonhomogeneous equation $T(\mathbf{x}) = \mathbf{y}$ is the sum*

$$\mathbf{x} = \mathbf{x}_p + N(T), \tag{6.8}$$

where $\mathbf{x}_p$ is a particular solution of $T(\mathbf{x}) = \mathbf{y}$ and $N(T)$ is the null space of T—the general solution of the homogeneous equation $T(\mathbf{x}) = \mathbf{0}$.

Proof Let $\mathbf{x} = \mathbf{x}_p + \mathbf{n}$ where $\mathbf{n}$ is some element of $N(T)$. By definition, $T(\mathbf{x}_p) = \mathbf{y}$ and $T(\mathbf{n}) = \mathbf{0}$. Since T is linear, $T(\mathbf{x}) = T(\mathbf{x}_p + \mathbf{n}) = T(\mathbf{x}_p) + T(\mathbf{n}) = \mathbf{y} + \mathbf{0} = \mathbf{y}$. This shows that every member of the set (6.8) is a solution of $T(\mathbf{x}) = \mathbf{y}$.

Now let $\mathbf{x}$ be an arbitrary solution of $T(\mathbf{x}) = \mathbf{y}$. By linearity of T, we have $T(\mathbf{x} - \mathbf{x}_p) = T(\mathbf{x}) - T(\mathbf{x}_p) = \mathbf{y} - \mathbf{y} = \mathbf{0}$. This means that $\mathbf{x} - \mathbf{x}_p$ is in the null space $N(T)$ which shows that every solution of $T(\mathbf{x}) = \mathbf{y}$ is contained in the set (6.8). Equation (6.8) is therefore the general solution of $T(\mathbf{x}) = \mathbf{y}$, as claimed. □

The structure theorem is an algorithm for solving a linear nonhomogeneous equation: find a particular solution of the nonhomogeneous equation—any solution will do—and add the general solution of the (usually simpler) homogeneous equation. In the context of ODE the general solution of the homogeneous equation is called *complimentary function* because it "compliments" the particular solution of the nonhomogeneous equation to its general solution. We will denote complimentary functions with the subscript "c" and write Eq. (6.8) as $\mathbf{x} = \mathbf{x}_p + \mathbf{x}_c$.

As the first example of using Theorem 6.1, let us apply it to Eq. (3.2) from Chap. 3. The operational form of that ODE is $T(x) = b$ where $T : x \mapsto \frac{dx}{dt} - ax$. This is the same transformation as the one that defines the natural growth equation: hence it is linear and the complimentary function is $x_c = C e^{at}$. As a particular solution we can take any function mapped into the constant function $y = b$ by T: the simplest choice is $x_p = -b/a$. According to Theorem 6.1, the general solution of (3.2) is

$$x = -\frac{b}{a} + C e^{at}.$$

This can be checked by imposing the initial condition $x(0) = x_0$ and comparing the result with (3.3).

Now, consider the second order linear ODE

$$\frac{d^2x}{dt^2} = t, \qquad (6.9)$$

which could be an instance of Newton's Second Law describing one-dimensional motion of a particle under the influence of a force that linearly increases with time. Integrating the right-hand side twice gives $x_p = t^3/6$. The general solution of the homogeneous equation

$$\frac{d^2x}{dt^2} = 0$$

can also be found through double indefinite integration: $x_c = C_1 + C_2 t$. According to Theorem 6.1, the general solution of (6.9) is therefore

$$x = x_p + x_c = \frac{t^3}{6} + C_1 + C_2 t.$$

The two indeterminate constants in x_c require two initial values: in the context of Newton's Second Law these are initial position and initial velocity of the particle.

The next two examples emphasize that Theorem 6.1 is a statement not just about linear ODE, but about all linear equations.

Consider the under-determined linear system

$$\begin{bmatrix} 1 & 1 & 1 \\ 1 & 2 & 3 \end{bmatrix} \begin{bmatrix} x_1 \\ x_2 \\ x_3 \end{bmatrix} = \begin{bmatrix} 3 \\ 6 \end{bmatrix} \qquad (6.10)$$

which has a particular solution $\mathbf{x}_p = \begin{bmatrix} 1 & 1 & 1 \end{bmatrix}^T$. To find $\mathbf{x}_c$, set $x_3 = C$ in

$$\begin{bmatrix} 1 & 1 & 1 \\ 1 & 2 & 3 \end{bmatrix} \begin{bmatrix} x_1 \\ x_2 \\ x_3 \end{bmatrix} = \begin{bmatrix} 0 \\ 0 \end{bmatrix}$$

and solve for x_1 and x_2. This gives $x_1 = C$ and $x_2 = -2C$, so the null space of the matrix is $\mathbf{x}_c = \begin{bmatrix} C & -2C & C \end{bmatrix}^T$. According to Theorem 6.6, the general solution of (6.10) is given by

$$\mathbf{x} = \mathbf{x}_p + \mathbf{x}_c = \begin{bmatrix} 1 \\ 1 \\ 1 \end{bmatrix} + \begin{bmatrix} C \\ -2C \\ C \end{bmatrix} = \begin{bmatrix} 1 \\ 1 \\ 1 \end{bmatrix} + C \begin{bmatrix} 1 \\ -2 \\ 1 \end{bmatrix}.$$

Notice how similar this is in form to the general solution of ODE (3.2) found in the first example.

Next, consider the Volterra equation (6.6) which, as we pointed out in Sect. 6.5, has the unique solution $P = P_0 e^{at}$. Differentiating the homogeneous equation

6.7 RC-Circuit Driven by a Simple Harmonic

$$P - a \int_0^t P(s)\,ds = 0 \tag{6.11}$$

with respect to time gives the natural growth equation $dP/dt - aP = 0$, while evaluating it at zero shows that $P(0) = 0$. The general solution of (6.11) is therefore the same as the solution of the natural growth ODE with the zero initial condition, which is the zero function. Since the complimentary function is zero, the general solution of the nonhomogeneous equation (6.6) is just the particular solution which is the same as the solution of the natural growth equation with $P(0) = P_0$.

It will be instructive to conclude this section with a counterexample. The ODE

$$\frac{dx}{dt} + x^2 = 1 + t^2 \tag{6.12}$$

has a particular solution $x_p = t$, however it is nonlinear and Theorem 6.1 does not apply. Although we know a particular solution, there is not much that we can do with it; we certainly cannot complete it to the general solution of (6.12) by adding the general solution of the "homogenous" ODE

$$\frac{dx}{dt} + x^2 = 0,$$

which is $x_c = (t + C)^{-1}$. There is a way to find the general solution of (6.12) but it requires specialized Calculus techniques which we describe at the end of the chapter.

6.7 RC-Circuit Driven by a Simple Harmonic

As a practical application of the structure theorem, let us analyse voltage V in the circuit shown in Fig. 6.1.

This is the same circuit as on the right side of Fig. 3.3 but with the switch removed and the battery replaced with a sinusoidal voltage source. Accordingly, the ODE for V is Eq. (3.14) derived in Sect. 3.3 but with E changed from a constant to a simple harmonic:

$$\frac{dV}{dt} + \frac{1}{RC} V = \frac{1}{RC} \left(a\,\cos(\omega t) + b\,\sin(\omega t) \right). \tag{6.13}$$

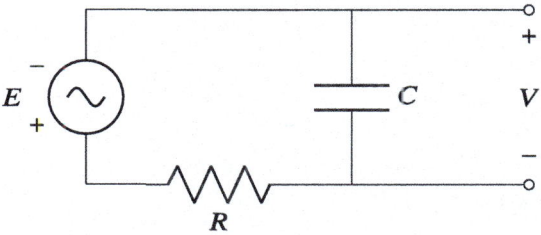

Fig. 6.1 *RC*-circuit driven by a sinusoidal voltage source

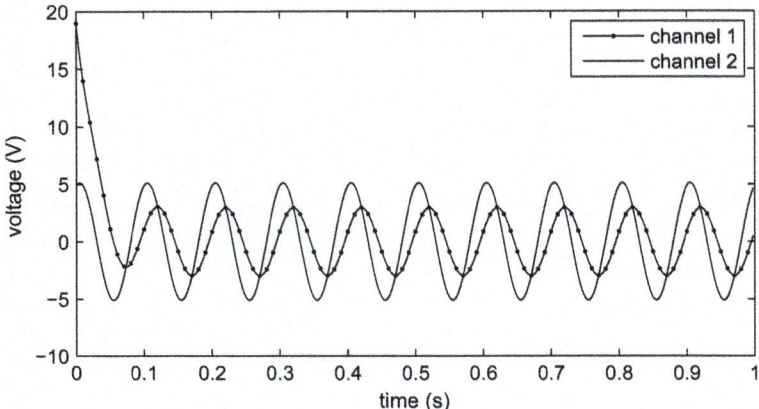

Fig. 6.2 Voltage V (channel 1) and source voltage E (channel 2) for the circuit in Fig. 6.1

The homogeneous ODE corresponding to (6.13)

$$\frac{dV}{dt} + \frac{1}{RC} V = 0,$$

has general solution $V_c = B\,e^{-\frac{t}{RC}}$ where we are forced to use B for the constant of integration because the traditional C is used for capacitance. With the complimentary function in place, we now require a particular solution. This is where examining experimental data can be very helpful.

Figure 6.2 shows V (channel 1) and $E = a\,\cos(\omega t) + b\,\sin(\omega t)$ (channel 2) measured with Digilent Analog Discovery™ oscilloscope. We used the same 2.2 µF capacitor (charged to about 20 V) and 10 kΩ resistor as in Sect. 3.3. The voltage source E was configured as a sine wave with frequency 10 Hz and amplitude 5 V. The measurements were taken for 1 s at the rate of 1 kHz; the dots on the trace for channel 1 show every tenth measurement.

It is evident from Fig. 6.2 that, after a transitional period, V settles to a simple harmonic of the same frequency as the source voltage. It is therefore sensible to guess that $V_p = \alpha\,\cos(\omega t) + \beta\,\sin(\omega t)$. To find the coefficients α and β, we substitute the guess for V_p into (6.13). On the left-hand side we get

$$\frac{dV}{dt} + \frac{1}{RC} V = \omega\,(-\alpha\,\sin(\omega t) + \beta\,\cos(\omega t)) + \frac{1}{RC}\,(\alpha\,\cos(\omega t) + \beta\,\sin(\omega t))$$
$$= \left(\frac{1}{RC}\alpha + \omega\beta\right)\cos(\omega t) + \left(-\omega\alpha + \frac{1}{RC}\beta\right)\sin(\omega t).$$

This must match the right-hand side $\frac{1}{RC}\,(a\,\cos(\omega t) + b\,\sin(\omega t))$. Equating like coefficients gives a linear system

6.7 RC-Circuit Driven by a Simple Harmonic

$$\frac{1}{RC}\alpha + \omega\beta = \frac{1}{RC}a, \quad -\omega\alpha + \frac{1}{RC}\beta = \frac{1}{RC}b,$$

whose solution is

$$\alpha = \frac{a - RC\omega b}{1 + (RC\omega)^2}, \quad \beta = \frac{RC\omega a + b}{1 + (RC\omega)^2}. \tag{6.14}$$

It follows that the general solution of ODE (6.13) is given by

$$V = \frac{(a - RC\omega b)\cos(\omega t) + (RC\omega a + b)\sin(\omega t)}{1 + (RC\omega)^2} + B\,e^{-\frac{t}{RC}}. \tag{6.15}$$

For a generic initial condition the constant of integration B should be set to

$$B = V_0 - \alpha = V_0 - \frac{a - RC\omega b}{1 + (RC\omega)^2}. \tag{6.16}$$

Figure 6.3 shows (6.15) (with B given by (6.16)) fitted to the data. To avoid clutter, we plotted every tenth data point as a circle, however we used the entire data set in the computation.

Equations (6.15) and (6.16) contain five parameters: a, b, ω, RC, and V_0. However, the last three of them are known: the circular frequency corresponding to 10 Hz sine wave E is $\omega = 20\pi$ rad/s; the time constant of the circuit was estimated in Sect. 3.3 as $RC = 22.753927046605931$ ms; and the initial condition is just the first observation which happens to be $V_0 = 18.970706242572529$ V.

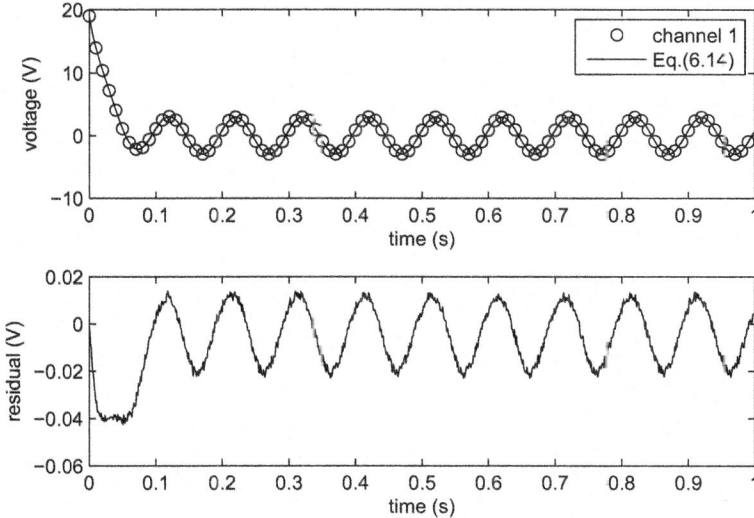

Fig. 6.3 The top panel shows (6.15) (with initial condition (6.16)) fitted to the data in Fig. 6.2; the bottom panel shows the residual

This leaves a and b which we estimated as

$$a = 4.853102024650312 \text{ V},$$
$$b = 1.564534740619342 \text{ V},$$

by fitting $a \cos(\omega t) + b \sin(\omega t)$ to the data from channel 2 using simple linear regression. Theoretically, a better estimate of a and b can be obtained by using the data from both channels. We did not pursue that because (i) the fit in Fig. 6.3 is satisfactory and (ii) the shape of the residual indicates that the model (6.13) does not fully explain the data. The latter is not a surprise: as we explained in Sect. 3.3, oscilloscope measurements are affected by probe loading, nonlinearities of circuit elements, and numerous other factors which a more advanced model must take into account.

In summary, we reaffirmed the structure theorem by showing that the response of a linear RC-circuit to a simple harmonic input is the sum of a simple harmonic of the same frequency, with coefficients linearly related to those of the input, and a decaying exponential, which only depends on the circuit: $V = V_p + V_c$. In Circuits, the complimentary function V_c is called a *transient* because it quickly decays and is only noticeable during the transition to "steady state" V_p.

6.8 Principle of Superposition

Let $T(\mathbf{x}) = \mathbf{y}$ be a linear equation. The principle of superposition is used when $\mathbf{y}$ is a linear combination (superposition) of vectors.

Theorem 6.2 (Principle of superposition) *Suppose that $\mathbf{x}_k$ is a solution of the linear equation $T(\mathbf{x}) = \mathbf{y}_k$ for $k = 1, \ldots, n$. Then $\sum_{k=1}^{n} c_k \mathbf{x}_k$ is a solution of $T(\mathbf{x}) = \sum_{k=1}^{n} c_k \mathbf{y}_k$.*

Proof By assumption, $T(\mathbf{x}_k) = \mathbf{y}_k$. Since T is linear,

$$T\left(\sum_{k=1}^{n} c_k \mathbf{x}_k\right) = \sum_{k=1}^{n} c_k T(\mathbf{x}_k) = \sum_{k=1}^{n} c_k \mathbf{y}_k,$$

which completes the proof. □

For all its simplicity, the principle of superposition is fundamentally important. As we will see in Chaps. 9 and 10, it allows the solutions of linear ODE with constant coefficients to be expressed as Fourier series and is at the core of *frequency response analysis*. In preparation for that, let us consider the circuit in Fig. 6.1 once again, but let us imagine that instead of generating a simple harmonic voltage the source E outputs a linear combination of harmonic voltages of different frequencies:

6.8 Principle of Superposition

$$E = \sum_{k=1}^{n} a_k \cos(\omega_k t) + b_k \sin(\omega_k t). \tag{6.17}$$

The voltage V then satisfies

$$\frac{dV}{dt} + \frac{1}{RC} V = \frac{1}{RC} \sum_{k=1}^{n} a_k \cos(\omega_k t) + b_k \sin(\omega_k t). \tag{6.18}$$

Equation (6.18) looks formidable. However, thanks to the principle of superposition and the results of Sect. 6.7, it is actually easy to solve. The complimentary function for (6.18) is the same as that for (6.13): $V_c = B\,e^{-\frac{t}{RC}}$

Thus the main challenge is to find a particular solution. According to Theorem 6.2, the latter can be found by adding particular solutions of

$$\frac{dV}{dt} + \frac{1}{RC} V = \frac{1}{RC} (a_k \cos(\omega_k t) + b_k \sin(\omega_k t)), \quad k = 1, \ldots, n,$$

which are simple harmonics $V_{p,k} = \alpha_k \cos(\omega_k t) + \beta_k \sin(\omega_k t)$ with coefficients given by (6.14):

$$\alpha_k = \frac{a_k - RC\,\omega_k\,b_k}{1 + (RC\,\omega_k)^2}, \quad \beta_k = \frac{RC\,\omega\,a_k + b_k}{1 + (RC\,\omega_k)^2}.$$

A particular solution of (6.18) is therefore

$$V_p = \sum_{k=1}^{n} \frac{(a_k - RC\,\omega_k\,b_k)\cos(\omega_k t) + (RC\,\omega_k\,a_k + b_k)\sin(\omega_k t)}{1 + (RC\,\omega_k)^2}. \tag{6.19}$$

In effect, the principle of superposition allows us to replace ODE (6.18) with a much simpler ODE (6.13).

We will fit (6.19) to experimental data in Chap. 9. In the meantime, let us validate it in MATLAB.

The similarly structured functions

```
function E = source(t,N)
E = zeros(size(t));
for k=1:N
    w = 20*pi*(2*k-1);
    a = 0; b = 100*pi/w;
    E = E + a*cos(w*t) + b*sin(w*t);
end
end
```

and

```
function Vp = particular(t,N)
Vp = zeros(size(t));
```

```
for n=1:N
    w     = 20*pi*(2*n-1);
    a     = 0; b = 100*pi/w;
    alpha = (a-RC*w*b)/(1+(RC*w)^2);
    beta  = (RC*w*a+b)/(1+(RC*w)^2);
    Vp    = Vp + alpha*cos(w*t) + beta*sin(w*t);
end
end
```

compute trigonometric sums (6.17) and (6.19), respectively. We set the frequencies $\omega_k = 20\pi(2k-1)$, and the coefficients $a_k = 0$ and $b_k = 100\pi/\omega_k$ in (6.17) so that `source(t,1)` is a 10 Hz sine wave with amplitude 5, similar to the source voltage in Sect. 6.7, and to make an interesting pattern emerge as the number of sine terms is increased. The computation also involves parameters RC and N. To make these parameters accessible to `source` and `particular` we placed the functions as subfunctions inside a main function block where we set $RC = 0.022$, a simple value close to the experimental RC-constant in Sect. 6.6; the number of harmonics N was varied as 1, 2, 4, 8.

To solve ODE (6.18) we use

```
odefun = @(t,V) (source(t,N)-V)/RC;
[t,V]  = ode45(odefun,linspace(0,.5,1000),20);
```

and, once the numerical solution is computed, the particular solution and the complimentary function are computed as follows:

```
Vp = particular(t,N);
Vc = (V(1)-Vp(1))*exp(-t/RC);
```

The particular solution is computed first because it is used to find the constant of integration in the complimentary function as $V_0 - V_p(0)$.

Figure 6.4 compares the numerical solution of (6.18) (circles) with $V_p + V_c$ (solid line) for $N = 1, 2, 4, 8$; the lines without circles are plots of (6.17).

Notice the similarity between the top left plot in Fig. 6.4 and the top plot in Fig. 6.3. Also notice how, as N increases, the plot of (6.17) begins to resemble a square waveform. As we will see in Chap. 9, that is not accidental.

6.9 Comments and Bibliography

Most ODE textbooks that we know of do not define vector spaces, and none define number fields. It is possible—and easy—to teach students to recognize linear ODE by sight and solve them using integrating factors, the method of undetermined coefficients, power series expansions, and other standard techniques. However, this is not our goal. We want our readers to learn not just the tactics for solving linear ODE, but also the strategy. That requires linear algebra in all of its abstract splendor.

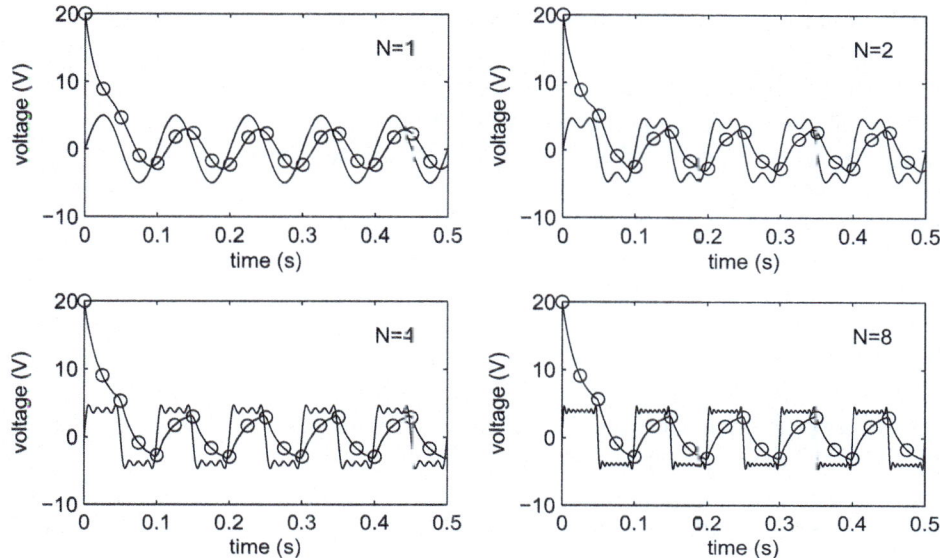

Fig. 6.4 Validation of (6.19) with $\omega_k = 20\pi(2k-1)$, $a_k = 0$, $b_k = 100\pi/\omega_k$, and $RC = 0.022$. The circles mark the numerical solution of (5.18) with $V_0 = 20$; the lines through the circles are plots of $V_p + V_c$ with V_p given by (6.19); the lines without circles are plots of (6.17)

As we explained in Sects. 6.1 and 6.2, axiomatic definitions of number fields and vector spaces are necessary for proper conceptualization of complex numbers and functional vector spaces. In our experience, students are often willing to suspend their disbelief when it comes to Euler's formula and the notion that functions are vectors. Yet, suspension of disbelief should not be the primary educational goal. Proper understanding of complex numbers and functional vector spaces is essential for understanding physics and engineering. The latter is impossible without firm understanding of how numbers and vectors are defined.

In Sect. 6.1 we described the progression

$$\mathbb{Z} \subset \mathbb{Q} \subset \mathbb{R} \subset \mathbb{C}$$

from integers to complex numbers. That may have made some readers wonder if that progression continues to a number system larger than the complex numbers. In a certain sense, it does, but what lies beyond $\mathbb{C}$ is the algebra of quaternions $\mathbb{H}$ which is not a number field.

Quaternions, discovered by Hamilton after more than a decade of intense searching, have three imaginary units i, j, and k subject to the relations

$$i^2 = j^2 = k^2 = ijk = -1.$$

Hamilton found these relations on October 16th, 1843 while on a walk along the Royal Canal in Dublin. He got so excited that he carved them on Broome Bridge (unfortunately, the carving is no longer visible but there is a commemorative plaque in its place). Addition and multiplication of quaternions (using the above relations) is defined on the set $\{a+bi+cj+dk \mid a,b,c,d \in \mathbb{R}\}$ in a manner similar to complex addition and multiplication. However, whereas complex addition and multiplication satisfy axioms (NF1)–(NF9), quaternion multiplication fails (NF6). While, technically, quaternions are not numbers, they are close in spirit and are useful for describing rotations in 3-space.

Hamilton's quaternions are followed by octonions $\mathbb{O}$ discovered the same year by his friend John T. Graves. Octonions have seven imaginary units and, whereas quaternions only fail commutativity of multiplication, octonions fail associativity of multiplication as well. There are no other *division algebras* over the real numbers following octonions.

Lest it appears from our exposition that the number universe is limited to $\mathbb{Q}$, $\mathbb{R}$, and $\mathbb{C}$, we should mention finite number fields which are constructed from integers using modular arithmetic; they are of great importance in computer science and we discuss them in one of the exercises below. Also, we would be remiss not to mention p-adic numbers which played a key role in the proof of Fermat's Last Theorem found by Andrew Wiles in 1995.

While defining vector spaces in Sect. 6.2 we gave a standard list of eight vector space axioms (VS1)–(VS8). That list is neither unique nor minimal. As shown in [4], defining a vector space over an arbitrary field requires six independent axioms; for certain fields that number can be reduced further.

In Sect. 6.3 we pointed out that changing the basis of a vector space changes the components of vectors. The manner in which vector components change makes them *rank*-1 *contravariant tensors*. Matrix representations of operators also change with the basis, but differently, in the way that makes them *rank*-2 *covariant tensors*.

In Sect. 6.4 we deferred an example of an isomorphism between number fields which we now provide. Consider the following mapping from complex numbers into the space of 2-by-2 real matrices:

$$\phi : a + bi \mapsto \begin{bmatrix} a & -b \\ b & a \end{bmatrix}$$

Let $z_1 = a_1 + b_1 i$ and $z_2 = a_2 + b_2 i$. As follows from the definition of ϕ and the rules of complex and matrix arithmetics,

$$\phi(z_1 + z_2) = \phi((a_1 + a_2) + (b_1 + b_2)i) = \begin{bmatrix} a_1 + a_2 & -(b_1 + b_2) \\ b_1 + b_2 & a_1 + a_2 \end{bmatrix}$$

$$= \begin{bmatrix} a_1 & -b_1 \\ b_1 & a_1 \end{bmatrix} + \begin{bmatrix} a_2 & -b_2 \\ b_2 & a_2 \end{bmatrix} = \phi(z_1) + \phi(z_2),$$

$$\phi(z_1 z_2) = \phi((a_1 a_2 - b_1 b_2) + (a_1 b_2 + b_1 a_2)i) = \begin{bmatrix} a_1 a_2 - b_1 b_2 & -(a_1 b_2 + b_1 a_2) \\ a_1 b_2 + b_1 a_2 & a_1 a_2 - b_1 b_2 \end{bmatrix}$$

$$= \begin{bmatrix} a_1 & -b_1 \\ b_1 & a_1 \end{bmatrix} \begin{bmatrix} a_2 & -b_2 \\ b_2 & a_2 \end{bmatrix} = \phi(z_1)\phi(z_2).$$

6.9 Comments and Bibliography

Thus ϕ preserves addition and multiplication and is therefore an isomorphism between $\mathbb{C}$ and a subspace of $M_{2\times 2}(\mathbb{R})$. Since two-by-two matrices can be thought of as operators on $\mathbb{R}^2$ we can think of complex numbers in the same way. The matrix corresponding to $z = a + bi$ represents dilation by $\sqrt{a^2 + b^2}$ and rotation by $\theta = \tan^{-1}(b/a)$:

$$\begin{bmatrix} a & -b \\ b & a \end{bmatrix} = \sqrt{a^2 + b^2} \begin{bmatrix} \cos(\theta) & -\sin(\theta) \\ \sin(\theta) & \cos(\theta) \end{bmatrix}$$

Thus a complex number encodes dilation and rotation of the plane. The number i, in particular, is a pure $90°$ rotation while its square is a $180°$ rotation: identifying complex numbers with matrices naturally resolves the $i^2 = -1$ "paradox."

In Sect. 6.5 we gave the general form of a scalar linear ODE (6.4) and its vector counterpart (6.5) to illustrate how to reduce both forms to

$$\frac{d\mathbf{u}}{dt} = M(t)\mathbf{u} + \mathbf{g}(t).$$

This reduction will be exploited in Chap. 7. Here we want to stress that one should not classify ODE as linear or nonlinear just by matching them against (6.4) and (6.5). It is far more useful, especially in the beginning stages, to test transformations for linearity and practice thinking in linear algebra terms. The significance of an ODE being linear is that its general solution has structure which invites natural questions: "How do I find a particular solution?", "What is the dimension of the null space?", "How do I construct a basis of the null space?" Matching an ODE against a linear ODE template does not facilitate inquiry and reasoning.

We should also caution that one must exercise a certain amount of care when isolating the transformation in a linear ODE. For instance, rewriting the linear homogeneous equation

$$\frac{d^2x}{dt^2} = -x$$

as

$$T(x) = \frac{1}{x}\frac{d^2x}{dt^2} = -1$$

ruins linearity.

At the end of Sect. 6.5 we mentioned Volterra integral equations which are equivalent to initial value problems. The Volterra equation (6.11), corresponding to IVP for the natural growth equation, can be solved as follows.

Let $T : P \to P_0 + a \int_0^t P(s)\,ds$. Then (6.11) may be written as $P = T(P)$ which shows that the solution is a *fixed point* of the transformation T. To find it, set $P_1 = 0$ and define $P_{n+1} = T(P_n)$. This gives the sequence

$$P_2 = P_0, \quad P_3 = P_0 + P_0\,a\,t, \quad P_4 = P_0 + P_0\,a\,t + P_0\frac{(a\,t)^2}{2}, \ldots$$

with the general term

$$P_n = P_0 \sum_{k=0}^{n-2} \frac{(a\,t)^k}{k!}$$

which approaches $P_0\, e^{at}$ as $n \to \infty$. This iterative scheme can be applied to a large class of Volterra equations. Computational details may be difficult but convergence—and therefore existence of the solution is assured by Brouwer's Fixed Point Theorem (proved in Real Analysis).

We stated in Sect. 6.6 that the general solution of a nonhomogeneous equation is not far removed from being a vector space. The technical term is *affine space*. Geometrically, the relation between vector spaces and affine spaces is the same as between lines (or planes) passing through the origin and general lines (or planes).

We ended Sect. 6.6 with a counterexample (6.12). This is an instance of a *Riccati equation*. The non-obvious substitution

$$x = \frac{1}{u}\frac{du}{dt}$$

transforms (6.12) into a second order linear homogeneous ODE

$$\frac{d^2 u}{dt^2} = (1 + t^2)\,u$$

which can be solved using power series. The general solution of (6.12) is

$$x = t + \frac{e^{-t^2}}{\int_0^t e^{-s^2}\,ds + C}.$$

It seems to have the structure $x_p + x_c$ but

$$\frac{e^{-t^2}}{\int_0^t e^{-s^2}\,ds + C}$$

is not connected to (6.12) in an obvious way.

To find V_p in Sect. 6.7 we used what some books call the method of undetermined coefficients and others call the guess-and-check method. We want to point out that if there was any guesswork in Sect. 6.7 it was informed by experimental data in Fig. 6.2. Instead of experimental data, which is not always available, one can equally well use numerical solution of the ODE. There is no need to blindly guess particular solutions. The checking part should also involve data, numerical solution, or, ideally, both.

We chose the example in Sect. 6.8 to show that a square wave—a signal familiar to anyone who studied Circuits—can be synthesized from simple harmonics. This raises the question: What other signals can be synthesized from simple harmonics? Remarkably, the answer is "All of them." There are some caveats, of course, but any physical signal may be represented as a superposition of harmonics. This all-important fact will be explained in Chaps. 8–10.

6.10 Exercises

We conclude with some general references for linear algebra. In previous chapters we cited textbooks that combine linear algebra with Calculus and ODE. They may be too advanced for beginners. Also, learning the linear algebra language and key ideas may be better from a single-purpose book, such as [3] or [5]. For readers wanting a much more thorough discussion of finite-dimensional vector spaces, we recommend Halmos's eponymous classic [1]. Halmos also has a famous linear algebra problem book [2] which is a great accompaniment to his vector space book.

6.10 Exercises

1. Let F be a number field. Use the number field axioms given in Sect. 6.1 to prove that $0 \times c = 0$ for every c in F.
2. Let $\oplus$ and $\otimes$ denote addition and multiplication of integers modulo 2. When reduced modulo 2 every integer is either 0 (if it is even) or 1 (if it is odd). We thus have a set $\{0, 1\}$ with the following addition and multiplication tables (Fig. 6.5): This is the finite number field $\mathbb{Z}_2$: you can check that (NF1)–(NF9) are satisfied. In the same way, using arithmetics modulo n, one defines $\mathbb{Z}_n$. Construct the tables of addition and multiplication for $\mathbb{Z}_3$, $\mathbb{Z}_4$, and $\mathbb{Z}_5$. Which of these are number fields?
3. Prove, using the axioms in Sect. 6.2, that a linear transformation $T : V \to W$ maps the zero vector in V into the zero vector in W.
4. Let $T : p \mapsto dp/dx + p$ be the transformation from the space of cubics P_3 into itself. Find the matrix representation of T using the basis of monomials $\{x^3, x^2, x, 1\}$ and confirm that its square corresponds to composition $T \circ T$.
5. Repeat the previous exercise using the basis

$$\left\{1, x, \frac{3}{2}x^2 - \frac{1}{2}, \frac{5}{2}x^3 - \frac{3}{2}x\right\}$$

6. Use the results of the preceding two problems to find a particular solution of

$$\frac{dp}{dx} + p = x^3$$

$\oplus$	0	1
0	0	1
1	1	0

$\otimes$	0	1
0	0	0
1	0	1

Fig. 6.5 Addition and multiplication in $\mathbb{Z}_2$

Fig. 6.6 Numerical solution of (6.20)

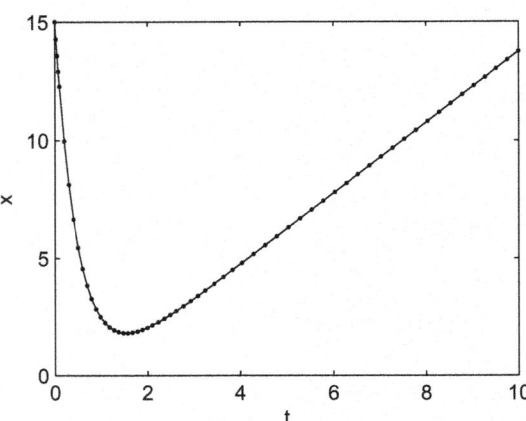

by converting the ODE into a matrix vector system $A\mathbf{c} = \mathbf{b}$ where $\mathbf{c}$ and $\mathbf{b}$ are the vectors of coefficients of p and x^3, respectively. Find the general solution as well and validate it in MATLAB using `ode45`.

7. The following code

```
[t,x] = ode45(@(t,x)3*t-1-2*x,[0 10],15);
plot(t,x,'k.-'); xlabel('t'); ylabel('x');
```

solves the IVP

$$\frac{dx}{dt} + 2x = 3t - 1, \quad x(0) = 15, \qquad (6.20)$$

and plots the solution shown in Fig. 6.6.

Use the figure as a guide to find the symbolic solution of (6.20). Better yet, run the code and analyse the data directly. Generalize the solution of (6.20) to the solution of

$$\frac{dx}{dt} + kx = at + b, \quad x(0) = x_0.$$

Validate your results in MATLAB.

8. Suppose that you can generate numerical solutions of

$$\frac{dx}{dt} = f(t, x), \quad x(0) = x_0 \qquad (6.21)$$

for different initial conditions x_0 on some interval, say, $[0, 1]$. Suppose further that computing solutions of (6.21) is very expensive. How can you test ODE (6.21) for linearity and what is the minimum number of numerical solutions required for that?

References

1. P.R. Halmos, *Finite-Dimensional Vector Spaces*. Undergraduate Texts in Mathematics. (Springer, 1993)
2. P.R. Halmos, *Linear Algebra Problem Book*. Number 16 in Dolciani Mathematical Expositions (Mathematical Association of America, Washington, DC, 1995)
3. C.D. Meyer, *Matrix Analysis and Applied Linear Algebra* (SIAM, 2001)
4. J.F. Rigby, J. Wiegold, Independent axioms for vector spaces. Math. Gazette **57**(399), 56–62 (1973)
5. G. Strang, *Linear Algebra and Its Applications*, 4th edn. (Brooks Cole, 2005)

Linear ODE with Constant Coefficients 7

In Chap. 6 we showed that every linear ODE can be cast into matrix-vector form

$$\frac{d\mathbf{u}}{dt} = M(t)\,\mathbf{u} + \mathbf{g}(t).$$

In this chapter we restrict our attention to linear ODE for which the matrix is constant: $M(t) = A$.

According to the structure theorem (Theorem 6.1 in Sect. 6.6), the general solution of

$$\frac{d\mathbf{u}}{dt} = A\,\mathbf{u} + \mathbf{g}(t) \tag{7.1}$$

is the sum $\mathbf{u}_p + \mathbf{u}_c$ where $\mathbf{u}_p$ is a particular solution of (7.1) and $\mathbf{u}_c$ is the general solution of the homogeneous ODE

$$\frac{d\mathbf{u}}{dt} = A\,\mathbf{u}. \tag{7.2}$$

Equation (7.2) was discussed in Sect. 5.4. As follows from that discussion, the general solution of (7.2)—the complimentary function—is $\mathbf{u}_c = e^{tA}\,\mathbf{c}$, where the matrix exponential is defined by the Taylor expansion (5.10) and $\mathbf{c}$ is a constant vector.

Summing up the Taylor series (5.10) directly is possible only for matrices with very simple structure. When the structure of A is not simple, computation of e^{tA} requires *eigendecomposition* which is introduced in Sect. 7.1. Eigendecomposition reduces the computation of e^{tA} to computation of e^{tD} where D is a diagonal matrix with *eigenvalues* of A on the main diagonal. Exponentiating diagonal matrices is a simple matter of exponentiating their main diagonals.

The computational examples in Sect. 7.1 come from Chap. 4: we use eigendecomposition to power a matrix of transition probabilities and solve a master equation from Sect. 4.4. These examples are reinforced with the discussion of a mass-spring system and its equivalent RLC-circuit in Sect. 7.2. The RLC-circuit is later studied experimentally in Sect. 7.7.

An immediate consequence of eigendecomposition is that general solutions of homogeneous linear ODE with constant coefficients are linear combinations of exponentials. We discuss that (and some important caveats) in Sect. 7.3 which concludes the discussion of homogeneous ODE.

The discussion of particular solutions of nonhomogeneous ODE is split between Sects. 7.4–7.6. If the components of **g** in (7.1) happen to be linear combinations of elementary functions—polynomials, harmonics, and exponentials—then the components of a particular solution $\mathbf{u}_p$ can be constructed as linear combinations of the same elementary functions with coefficients determined by solving linear algebraic systems: this is the method of undetermined coefficients which we have already used in Sect. 6.7 and now formally introduce in Sect. 7.4. If **g** is not elementary, then $\mathbf{u}_p$ can be computed using *convolution*. One-dimensional convolution formula is derived in Sect. 7.5 and then generalized to multiple dimensions in Sect. 7.6.

The final Sect. 7.7 shows how multidimensional convolution is used to derive input-output relations, such as the *impulse response*. As an illustration, we compute the impulse response of an RLC-circuit from the measurement of its step response.

7.1 Eigendecomposition

Let P be a nonsingular (invertible) matrix of the same size as A in (7.2). The linear change of variables $\mathbf{u} = P\mathbf{v}$ transforms (7.2) into

$$\frac{d\mathbf{v}}{dt} = P^{-1} A P \mathbf{v}. \tag{7.3}$$

If P is chosen strategically, the matrix $P^{-1} A P$ in (7.3) may be more amenable to exponentiation than the matrix A in (7.2). In an ideal scenario $P^{-1} A P = D$ where D is a diagonal matrix

$$D = \begin{bmatrix} \lambda_1 & & & 0 \\ & \lambda_2 & & \\ & & \ddots & \\ 0 & & & \lambda_n \end{bmatrix} = \mathrm{diag}(\lambda_1, \lambda_2, \ldots, \lambda_n).$$

In this case the system (7.3) decouples into n scalar ODE

$$\frac{dv_1}{dt} = \lambda_1 v_1, \quad \frac{dv_2}{dt} = \lambda_2 v_2, \quad \ldots, \quad \frac{dv_n}{dt} = \lambda_n v_n,$$

which can be solved independently:

$$v_1 = c_1 e^{\lambda_1 t}, \quad v_2 = c_2 e^{\lambda_2 t}, \quad \ldots, \quad v_n = c_n e^{\lambda_n t}.$$

7.1 Eigendecomposition

In matrix-vector language, if $P^{-1} A F = D$ in (7.3), then $\mathbf{v} = e^{tD} \mathbf{c}$, where $\mathbf{c}$ is a constant vector with components c_k and e^{tD} is the diagonal matrix with exponentials $e^{\lambda_k t}$ on the main diagonal:

$$e^{tD} = \begin{bmatrix} e^{\lambda_1 t} & & & 0 \\ & e^{\lambda_2 t} & & \\ & & \ddots & \\ 0 & & & e^{\lambda_n t} \end{bmatrix} = \text{diag}\left(e^{\lambda_1 t}, e^{\lambda_2 t}, \ldots, e^{\lambda_n t}\right). \tag{7.4}$$

Equation (7.4) can also be derived from (5.10); we leave that as an exercise.

Not every matrix is diagonalizable. However, let us assume that A is and let us find the matrix P. The factorization $P^{-1} A P = D$ is equivalent to

$$A P = P D. \tag{7.5}$$

Let $\mathbf{p}_k$ be the k-th column of P. As follows from the row-by-column rule of matrix multiplication, the multiplication of P by A on the left leads to each column of P being multiplied by A:

$$A P = A \begin{bmatrix} \mathbf{p}_1 & \mathbf{p}_2 & \cdots & \mathbf{p}_n \end{bmatrix} = \begin{bmatrix} A \mathbf{p}_1 & A \mathbf{p}_2 & \cdots & A \mathbf{p}_n \end{bmatrix}.$$

Meanwhile, multiplication of P by the diagonal matrix D on the right is the same as scaling the columns of P by the corresponding diagonal entries of D:

$$P D = \begin{bmatrix} \mathbf{p}_1 & \mathbf{p}_2 & \cdots & \mathbf{p}_n \end{bmatrix} \text{diag}(\lambda_1, \ldots, \lambda_n) = \begin{bmatrix} \lambda_1 \mathbf{p}_1 & \lambda_2 \mathbf{p}_2 & \cdots & \lambda_n \mathbf{p}_n \end{bmatrix}.$$

Equation (7.5) therefore states that

$$\begin{bmatrix} A \mathbf{p}_1 & A \mathbf{p}_2 & \cdots & A \mathbf{p}_n \end{bmatrix} = \begin{bmatrix} \lambda_1 \mathbf{p}_1 & \lambda_2 \mathbf{p}_2 & \cdots & \lambda_n \mathbf{p}_n \end{bmatrix},$$

whence follows that $A \mathbf{p}_k = \lambda_k \mathbf{p}_k$ for $k = 1, \ldots, n$. In words, multiplication by A has the same effect on the columns of P as scaling. We formalize this as the following definition.

Definition 7.1 A nonzero vector $\mathbf{p}$ is an *eigenvector* of a square matrix A with corresponding *eigenvalue* λ if $A \mathbf{p} = \lambda \mathbf{p}$.

In the language of Definition 7.1, the columns of P are eigenvectors of A and the diagonal elements of D are the corresponding eigenvalues. Henceforth we will refer to both the pair of matrices $\{P, D\}$ and the factorization $A = P D P^{-1}$ as the *eigendecomposition* of A.

Symbolic computation of eigendecomposition begins with the computation of eigenvalues. According to Definition 7.1, $\mathbf{p}$ is a nonzero solution of the homogeneous system $(A - \lambda I) \mathbf{p} = \mathbf{0}$ where $I = \text{diag}(1, 1, \ldots, 1)$ is the identity matrix. Now, a square homogeneous system possesses a nontrivial solution if and only if its matrix is singular (non-invertible). Therefore, in order to ensure the existence of eigenvectors, we must make the matrix $A - \lambda I$ singular. That can only be accomplished by choosing λ so that

$$\det(A - \lambda I) = 0. \tag{7.6}$$

As a function of λ, the determinant in (7.6) is a polynomial of degree n, called the *characteristic polynomial* of A. According to the Fundamental Theorem of Algebra, a polynomial of degree n has n complex roots. Once the roots of (7.6)—the characteristic roots λ_k—are found, the corresponding eigenvectors $\mathbf{p}_k$ are computed as nonzero solutions of $(A - \lambda_k I)\,\mathbf{p}_k = \mathbf{0}$. These solutions are not unique: an eigenvector can be scaled to produce another eigenvector with the same eigenvalue.

As the first example, let us compute the eigendecomposition of

$$\begin{bmatrix} e^{-k_1 \Delta t} & 1 - e^{-k_2 \Delta t} \\ 1 - e^{-k_1 \Delta t} & e^{-k_2 \Delta t} \end{bmatrix}. \tag{7.7}$$

We encountered this matrix in Sect. 4.4—it is the matrix of transition probabilities of the Markov model of mutarotation of glucose. The characteristic polynomial

$$\det\left(\begin{bmatrix} e^{-k_1 \Delta t} & 1 - e^{-k_2 \Delta t} \\ 1 - e^{-k_1 \Delta t} & e^{-k_2 \Delta t} \end{bmatrix} - \lambda \begin{bmatrix} 1 & 0 \\ 0 & 1 \end{bmatrix}\right) = \det\left(\begin{bmatrix} e^{-k_1 \Delta t} - \lambda & 1 - e^{-k_2 \Delta t} \\ 1 - e^{-k_1 \Delta t} & e^{-k_2 \Delta t} - \lambda \end{bmatrix}\right)$$

$$= \lambda^2 - \left(e^{-k_1 \Delta t} + e^{-k_1 \Delta t}\right)\lambda + e^{-k_1 \Delta t} + e^{-k_2 \Delta t} - 1$$

has zeros

$$\lambda_1 = e^{-k_1 \Delta t} + e^{-k_2 \Delta t} - 1, \quad \lambda_2 = 1.$$

These are the eigenvalues. The first eigenvector, corresponding to λ_1, is any nontrivial solution of

$$\left(\begin{bmatrix} e^{-k_1 \Delta t} & 1 - e^{-k_2 \Delta t} \\ 1 - e^{-k_1 \Delta t} & e^{-k_2 \Delta t} \end{bmatrix} - \left(e^{-k_1 \Delta t} + e^{-k_2 \Delta t} - 1\right)\begin{bmatrix} 1 & 0 \\ 0 & 1 \end{bmatrix}\right)\begin{bmatrix} x \\ y \end{bmatrix} = \begin{bmatrix} 0 \\ 0 \end{bmatrix}$$

which simplifies to

$$\begin{bmatrix} 1 - e^{-k_2 \Delta t} & 1 - e^{-k_2 \Delta t} \\ 1 - e^{-k_1 \Delta t} & 1 - e^{-k_1 \Delta t} \end{bmatrix}\begin{bmatrix} x \\ y \end{bmatrix} = \begin{bmatrix} 0 \\ 0 \end{bmatrix},$$

or, in scalar form,

$$\left(1 - e^{-k_2 \Delta t}\right)(x + y) = 0, \quad \left(1 - e^{-k_1 \Delta t}\right)(x + y) = 0.$$

The two equations are linearly dependent (as they should be) and are equivalent to $x + y = 0$. We can therefore set $\mathbf{p}_1 = C\,\begin{bmatrix} 1 & -1 \end{bmatrix}^T$ where C is any nonzero constant; while the human choice is $C = 1$, software routines, such as MATLAB's eig, normalize eigenvectors to have unit length and therefore would set $C = 1/\sqrt{2}$ in this case. We leave it as an exercise to show

7.1 Eigendecomposition

that the second eigenvector, corresponding to λ_2, is $\mathbf{p}_2 = C \begin{bmatrix} 1 - e^{-k_2 \Delta t} & 1 - e^{-k_1 \Delta t} \end{bmatrix}^T$ where, again, we may set $C = 1$, for simplicity. The eigendecomposition of (7.7) is therefore

$$P = \begin{bmatrix} 1 & 1 - e^{-k_2 \Delta t} \\ -1 & 1 - e^{-k_1 \Delta t} \end{bmatrix} \quad D = \begin{bmatrix} e^{-k_1 \Delta t} + e^{-k_2 \Delta t} - 1 & 0 \\ 0 & 1 \end{bmatrix}. \quad (7.8)$$

In Sect. 4.4 we did not power the matrix (7.7) because its powers do not exhibit a simple pattern. Having found the eigendecomposition (7.8), we no longer need to search for a pattern. Since

$$(P D P^{-1})^n = P D \underbrace{P^{-1} P}_{I} D \underbrace{P^{-1} P}_{I} D P^{-1} \ldots P D \underbrace{P^{-1} P}_{I} D P^{-1}$$

$$= P D^n P^{-1}$$

we just have to power the diagonal matrix D which amounts to powering the eigenvalues on its main diagonal. After carrying out triple matrix multiplication

$$\begin{bmatrix} 1 & 1 - e^{-k_2 \Delta t} \\ -1 & 1 - e^{-k_1 \Delta t} \end{bmatrix} \begin{bmatrix} \left(e^{-k_1 \Delta t} + e^{-k_2 \Delta t} - 1\right)^n & 0 \\ 0 & 1 \end{bmatrix} \begin{bmatrix} 1 & 1 - e^{-k_2 \Delta t} \\ -1 & 1 - e^{-k_1 \Delta t} \end{bmatrix}^{-1}$$

and simplifying the result, we get

$$\begin{bmatrix} e^{-k_1 \Delta t} & 1 - e^{-k_2 \Delta t} \\ 1 - e^{-k_1 \Delta t} & e^{-k_2 \Delta t} \end{bmatrix}^n = \frac{\left(e^{-k_1 \Delta t} + e^{-k_2 \Delta t} - 1\right)^n}{2 - e^{-k_1 \Delta t} - e^{-k_2 \Delta t}} \begin{bmatrix} a & b \\ c & d \end{bmatrix}$$

where

$$a = 1 - e^{-k_1 \Delta t} + \frac{1 - e^{-k_2 \Delta t}}{\left(e^{-k_1 \Delta t} + e^{-k_2 \Delta t} - 1\right)^n},$$

$$b = e^{-k_2 \Delta t} - 1,$$

$$c = e^{-k_1 \Delta t} - 1,$$

$$d = 1 - e^{-k_2 \Delta t} + \frac{1 - e^{-k_1 \Delta t}}{\left(e^{-k_1 \Delta t} + e^{-k_2 \Delta t} - 1\right)^n}.$$

In Sect. 4.4, rather than power (7.7) we used it to derive Kolmogorov's forward equation (4.13)

$$\frac{d}{dt} \begin{bmatrix} x \\ y \end{bmatrix} = \begin{bmatrix} -k_1 & k_2 \\ k_1 & -k_2 \end{bmatrix} \begin{bmatrix} x \\ y \end{bmatrix},$$

which we solved by splitting into components and eliminating y using mass balance. We will now solve this ODE directly, using matrix exponentiation.

We leave it as an exercise to show that the eigendecomposition of the matrix of (4.13) is given by

$$P = \begin{bmatrix} 1 & k_2 \\ -1 & k_1 \end{bmatrix}, \quad D = \begin{bmatrix} -(k_1 + k_2) & 0 \\ 0 & 0 \end{bmatrix}.$$

Setting $X = t\,P\,D\,P^{-1}$ in (5.10) results in

$$e^{t\,P\,D\,P^{-1}} = \sum_{k=0}^{\infty} \frac{t^k}{k!}\left(P\,D\,P^{-1}\right)^k = \sum_{k=0}^{\infty} \frac{t^k}{k!}\,P\,D^k\,P^{-1} = P\left(\sum_{k=0}^{\infty} \frac{t^k}{k!}\,D^k\right)P^{-1}$$
$$= P\,e^{t\,D}\,P^{-1}.$$

Therefore

$$\exp\left(t\begin{bmatrix}-k_1 & k_2 \\ k_1 & -k_2\end{bmatrix}\right) = \begin{bmatrix}1 & k_2 \\ -1 & k_1\end{bmatrix}\begin{bmatrix}e^{-(k_1+k_2)t} & 0 \\ 0 & 1\end{bmatrix}\begin{bmatrix}1 & k_2 \\ -1 & k_1\end{bmatrix}^{-1}$$
$$= \frac{1}{k_1+k_2}\begin{bmatrix}k_1\,e^{-(k_1+k_2)t} + k_2 & k_2\left(1 - e^{-(k_1+k_2)t}\right) \\ k_1\left(1 - e^{-(k_1+k_2)t}\right) & k_2\,e^{-(k_1+k_2)t} + k_1\end{bmatrix}$$

and hence the solution of (4.13) is

$$\begin{bmatrix}x \\ y\end{bmatrix} = \frac{1}{k_1+k_2}\begin{bmatrix}k_1\,e^{-(k_1+k_2)t} + k_2 & k_2\left(1 - e^{-(k_1+k_2)t}\right) \\ k_1\left(1 - e^{-(k_1+k_2)t}\right) & k_2\,e^{-(k_1+k_2)t} + k_1\end{bmatrix}\begin{bmatrix}x_0 \\ y_0\end{bmatrix},$$

which matches our earlier solution (4.14).

7.2 1DOF Mass-Spring System and Its Equivalent Circuit

Figure 7.1 depicts a cart of mass m that is anchored to a wall with a spring of stiffness k.

The cart is constrained to move along the x-axis and thus has a single degree of freedom (1DOF). As is common in mechanics, we describe the motion using displacement from equilibrium x, shown with an arrow. It is also common in mechanics to denote time derivatives using dots placed over the functions. Accordingly, we will denote the velocity of the cart as $\dot{x}$ and its acceleration as $\ddot{x}$.

One of the forces acting on the cart is the spring force which we assume to be given by Hooke's law. Another force, not shown in the figure, is friction: this we assume to be

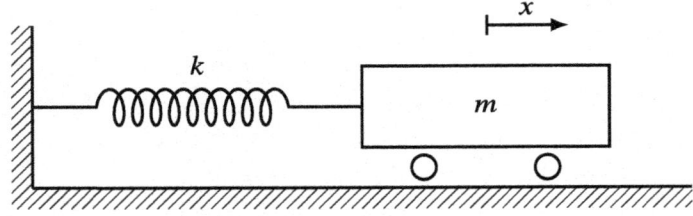

Fig. 7.1 Mass-spring system

7.2 1DOF Mass-Spring System and Its Equivalent Circuit

proportional to the cart's velocity. With these assumptions, Newton's Second Law translates into the following ODE

$$m\ddot{x} = -kx - r\dot{x}, \quad \left(\dot{\cdot} = \frac{d}{dt}\right). \tag{7.9}$$

Equation (7.9) is a second order linear homogeneous ODE with constant coefficients.

In order to convert (7.9) into matrix-vector form, we follow the procedure described in Sect. 6.5 and introduce $\mathbf{u} = \begin{bmatrix} x & \dot{x} \end{bmatrix}^T$. Then

$$\dot{\mathbf{u}} = \begin{bmatrix} \dot{x} \\ \ddot{x} \end{bmatrix} = \begin{bmatrix} \dot{x} \\ -\frac{k}{m}x - \frac{r}{m}\dot{x} \end{bmatrix} = \begin{bmatrix} 0 & 1 \\ -\frac{k}{m} & -\frac{r}{m} \end{bmatrix} \begin{bmatrix} x \\ \dot{x} \end{bmatrix} = A\,\mathbf{u}.$$

The matrix

$$A = \begin{bmatrix} 0 & 1 \\ -\frac{k}{m} & -\frac{r}{m} \end{bmatrix} \tag{7.10}$$

completely describes the mass-spring system in Fig. 7.1 and, from the mathematical point of view, is its most convenient description.

The characteristic polynomial of (7.10) has two roots given by the quadratic formula

$$\lambda_{1,2} = \frac{-r \pm \sqrt{r^2 - 4mk}}{2m}. \tag{7.11}$$

These are the eigenvalues—complex conjugates with negative real parts if $r^2 < 4mk$ or negative real numbers if $r^2 > 4mk$ (the case when $r^2 = 4mk$ will be addressed later). As the corresponding eigenvectors we can take $\mathbf{p}_j = \begin{bmatrix} 1 & \lambda_j \end{bmatrix}^T$, $j = 1, 2$. Indeed, since $\lambda_j^2 + \frac{r}{m}\lambda_j + \frac{k}{m} = 0$, we have

$$\begin{bmatrix} 0 & 1 \\ -\frac{k}{m} & -\frac{r}{m} \end{bmatrix} \begin{bmatrix} 1 \\ \lambda_j \end{bmatrix} = \begin{bmatrix} \lambda_j \\ -\frac{k}{m} - \frac{r}{m}\lambda_j \end{bmatrix} = \begin{bmatrix} \lambda_j \\ \lambda_j^2 \end{bmatrix} = \lambda_j \begin{bmatrix} 1 \\ \lambda_j \end{bmatrix},$$

in accordance with Definition 7.1. The eigendecomposition of (7.10) is thus

$$P = \begin{bmatrix} 1 & 1 \\ \lambda_1 & \lambda_2 \end{bmatrix}, \quad D = \begin{bmatrix} \lambda_1 & 0 \\ 0 & \lambda_2 \end{bmatrix} \tag{7.12}$$

with λ's given by (7.11). As follows from (7.12),

$$\exp\left(t \begin{bmatrix} 0 & 1 \\ -\frac{k}{m} & -\frac{r}{m} \end{bmatrix}\right) = \begin{bmatrix} 1 & 1 \\ \lambda_1 & \lambda_2 \end{bmatrix} \begin{bmatrix} e^{\lambda_1 t} & 0 \\ 0 & e^{\lambda_2 t} \end{bmatrix} \begin{bmatrix} 1 & 1 \\ \lambda_1 & \lambda_2 \end{bmatrix}^{-1}$$
$$= \frac{1}{\lambda_2 - \lambda_1} \begin{bmatrix} \lambda_2 e^{\lambda_1 t} - \lambda_1 e^{\lambda_2 t} & e^{\lambda_2 t} - e^{\lambda_1 t} \\ \lambda_1 \lambda_2 \left(e^{\lambda_1 t} - e^{\lambda_2 t}\right) & \lambda_2 e^{\lambda_2 t} - \lambda_1 e^{\lambda_1 t} \end{bmatrix} \tag{7.13}$$

and therefore

$$\begin{bmatrix} x \\ \dot{x} \end{bmatrix} = \frac{1}{\lambda_2 - \lambda_1} \begin{bmatrix} \lambda_2 e^{\lambda_1 t} - \lambda_1 e^{\lambda_2 t} & e^{\lambda_2 t} - e^{\lambda_1 t} \\ \lambda_1 \lambda_2 \left(e^{\lambda_1 t} - e^{\lambda_2 t} \right) & \lambda_2 e^{\lambda_2 t} - \lambda_1 e^{\lambda_1 t} \end{bmatrix} \begin{bmatrix} x_0 \\ \dot{x}_0 \end{bmatrix}.$$

Carrying out matrix-vector multiplication gives the solution of (7.9) as

$$x = x_0 \frac{\lambda_2 e^{\lambda_1 t} - \lambda_1 e^{\lambda_2 t}}{\lambda_2 - \lambda_1} + \dot{x}_0 \frac{e^{\lambda_2 t} - e^{\lambda_1 t}}{\lambda_2 - \lambda_1}. \tag{7.14}$$

As stated, the eigendecomposition (7.12) requires the eigenvalues of (7.10) to be distinct. If $\lambda_1 = \lambda_2$ then the matrix P in (7.12) becomes singular which signifies that the matrix (7.10) is not diagonalizable. While this may seem to be a cause for concern, it really is not for two reasons. Firstly, there is absolutely no chance that a real mass-spring system will have parameters m, r, and k exactly such that $r^2 = 4mk$. Secondly, while (7.12) stops being a valid eigendecomposition when $\lambda_1 = \lambda_2$, the solution can still be interpreted as a limit. For instance, taking the limit of (7.14) as $\lambda_{1,2} \to \lambda$ gives

$$x = x_0 e^{\lambda t} (1 - \lambda t) + \dot{x}_0 t e^{\lambda t}, \tag{7.15}$$

which is a valid solution of (7.9) when $r^2 = 4mk$ and $\lambda = -r/2m$ is a repeated eigenvalue of multiplicity 2.

The mass-spring system in Fig. 7.1 has an electrical analog shown in Fig. 7.2.

This is an RLC-circuit—the RC-circuit of Sect. 3.3 with the addition of a new element, the *inductor*.

Physically, an inductor is a coil of wire, often wound around a ferromagnetic core. A changing current passing through a coil creates a magnetic field which resists the current that creates it. Continuing the hydrodynamic analogy of Sect. 3.3, we may liken an inductor to a paddle wheel. Imagine a paddle wheel spun by a steady current which is suddenly reversed. Due to inertia, it will take some time for the paddle wheel to change its direction of rotation while it is fighting against the change in the current.

Let I denote electric current. Since the RLC-circuit in Fig. 7.2 consists of a single loop, the current through all three elements is the same and, also, according to Kirchhoff's Voltage Law, the sum of voltage drops across the elements is zero:

Fig. 7.2 RLC-circuit

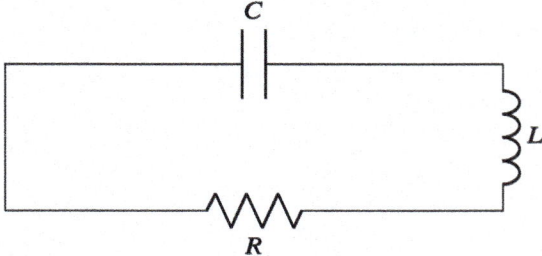

$$L\frac{dI}{dt} + RI + \frac{Q}{C} = 0. \tag{7.16}$$

The last two terms on the left-hand side of (7.16) are Ohm's Law and Volta's Law which we introduced in Sect. 3.3; the first term is the voltage drop for a linear inductor—a special case of *Faraday's Law of Induction*. Setting $I = dQ/dt$ in (7.16) gives a second order ODE of the same form as (7.9):

$$L\frac{d^2Q}{dt^2} + R\frac{dQ}{dt} + \frac{Q}{C} = 0. \tag{7.17}$$

In fact, if we replace capacitor's charge Q with displacement x, use the dot notation for time derivatives, and change the coefficients in (7.17) according to the scheme

$$L \leftrightarrow m, \quad R \leftrightarrow r, \quad \frac{1}{C} \leftrightarrow k, \tag{7.18}$$

then (7.17) becomes (7.9). The correspondence (7.18) is known as the *electro-mechanical analogy*. Just as a thermal circuit can be realized as an RC-circuit (Sect. 3.9), a mass-spring system can be realized as an RLC-circuit with inductors playing the role of masses and capacitors playing the role of springs.

7.3 General Solutions of Homogeneous ODE

In the preceding section, we solved a second order homogeneous ODE (7.9) by converting it into a first order vector ODE, finding, in the end, that the solution is a linear combination of two exponentials. Armed with this insight, we can solve (7.9) directly, as follows. Let us "guess" its particular solutions to be exponential. Substituting $x = e^{\lambda t}$ into (7.9) and canceling $e^{\lambda t}$ gives

$$m\lambda^2 + r\lambda + k = 0.$$

This is the characteristic polynomial of ODE (7.9): notice that it is a scalar multiple of the characteristic polynomial of the matrix (7.10). The characteristic roots are, of course, the same as the eigenvalues (7.11). If the characteristic roots are distinct, we have two linearly independent exponential solutions $e^{\lambda_1 t}$ and $e^{\lambda_2 t}$ whose superposition is, by homogeneity, also a solution—in fact, it is the general solution $x = C_1 e^{\lambda_1 t} + C_2 e^{\lambda_2 t}$. If the characteristic roots are repeated, $\lambda_1 = \lambda_2 = \lambda$, the method gives only one exponential solution $e^{\lambda t}$. According to (7.15), another independent solution is $t e^{\lambda t}$ and therefore the general solution for repeated roots is $x = C_1 e^{\lambda t} + C_2 t e^{\lambda t}$.

For a general homogeneous ODE of order n (with constant coefficients)

$$a_n x^{(n)} + a_{n-1} x^{(n-1)} + a_{n-2} x^{(n-2)} + \cdots + a_0 x = 0,$$

the characteristic polynomial

$$a_n \lambda^n + a_{n-1} \lambda^{n-1} + a_{n-2} \lambda^{n-2} + \cdots + a_0 = 0,$$

will typically have n distinct complex roots. If all of the characteristic roots λ_k are distinct then the general solution is $x = \sum_{k=1}^{n} C_k e^{\lambda_k t}$. If some of the roots are repeated, the corresponding exponentials should be multiplied by powers of t running from zero to one less than the multiplicity. For instance, the characteristic polynomial of

$$x^{(5)} - x^{(4)} - 2x^{(3)} + 2x^{(2)} + x^{(1)} - x = 0$$

has two roots: $\lambda_1 = 1$ with multiplicity 3 and $\lambda_2 = -1$ with multiplicity 2. The general solution is therefore

$$x = C_1 e^t + C_2 t e^t + C_3 t^2 e^t + C_4 e^{-t} + C_5 t e^{-t}.$$

When solving initial value problems it is best to convert higher order linear ODE with constant coefficients into first order matrix-vector systems. However, if only the general solution is needed, "guessing" exponential solutions is the quickest way to construct it.

7.4 Method of Undetermined Coefficients

It is time to turn our attention to nonhomogeneous equations. The differential operator with constant coefficients

$$x \mapsto a_n x^{(n)} + a_{n-1} x^{(n-1)} + a_{n-2} x^{(n-2)} + \cdots + a_0 x$$

maps polynomials into polynomials, exponentials into exponentials, and simple harmonics into simple harmonics. If the right-hand side of

$$a_n x^{(n)} + a_{n-1} x^{(n-1)} + a_{n-2} x^{(n-2)} + \cdots + a_0 x = f(t)$$

is a linear combination of these functions, the form of a particular solution can be guessed up to coefficients which can then be determined by substituting the guess into the ODE and forcing equality. Having already used this strategy in Sect. 6.7, we now officially introduce it as the method of undetermined coefficients.

As an example, let us find a particular solution of

$$m \ddot{x} + r \dot{x} + k x = a \cos(\omega t) + b \sin(\omega t). \tag{7.19}$$

This is the equation of motion for the mass-spring system in Fig. 7.1 with the addition of an external harmonic force. Setting $x = \alpha \cos(\omega t) + \beta \sin(\omega t)$ in (7.19) results in

7.4 Method of Undetermined Coefficients

$$m\ddot{x} + r\dot{x} + kx = -m\omega^2 \left(\alpha \cos(\omega t) + \beta \sin(\omega t)\right)$$
$$+ r\omega \left(-\alpha \sin(\omega t) - \beta \cos(\omega t)\right) + k \left(\alpha \cos(\omega t) + \beta \sin(\omega t)\right)$$
$$= \left((k - m\omega^2)\alpha + r\omega\beta\right)\cos(\omega t) + \left(-r\omega\alpha + (k - m\omega^2)\beta\right)\sin(\omega t)$$
$$= a\cos(\omega t) + b\sin(\omega t).$$

Hence α and β must satisfy

$$(k - m\omega^2)\alpha + r\omega\beta = a, \quad -r\omega\alpha + (k - m\omega^2)\beta = b.$$

Solving these linear equations gives

$$\alpha = \frac{(k - m\omega^2)a - r\omega b}{(k - m\omega^2)^2 + r^2\omega^2}, \quad \beta = \frac{r\omega a + (k - m\omega^2)b}{(k - m\omega^2)^2 - r^2\omega^2}.$$

Therefore, a particular solution of (7.19) is

$$x_p = \frac{(k - m\omega^2)a - r\omega b}{(k - m\omega^2)^2 + r^2\omega^2}\cos(\omega t) + \frac{r\omega a + (k - m\omega^2)b}{(k - m\omega^2)^2 + r^2\omega^2}\sin(\omega t). \tag{7.20}$$

Computation of (7.20) can be simplified through the use of complex trigonometry. Using identities (5.6) from Sect. 5.3, we can express a simple harmonic as a linear combination of conjugate complex exponentials:

$$a\cos(\omega t) + b\sin(\omega t) = \frac{a - bi}{2}e^{i\omega t} + \frac{a + bi}{2}e^{-i\omega t}.$$

This invites the use of the principle of superposition.

Setting $x = \gamma e^{i\omega t}$ in

$$m\ddot{x} + r\dot{x} + kx = e^{i\omega t} \tag{7.21}$$

and solving

$$m(i\omega)^2 \gamma e^{i\omega t} + r(i\omega)\gamma e^{i\omega t} + k\gamma e^{i\omega t} = e^{i\omega t}$$

for γ gives a particular solution of (7.21) as

$$x_p = \frac{1}{m(i\omega)^2 + r(i\omega) + k} e^{i\omega t}. \tag{7.22}$$

Notice that the denominator in (7.22) is the characteristic polynomial of (7.21) evaluated at $i\omega$. The computation is the same if the right-hand side of (7.21) is replaced with its conjugate—the result is the conjugate of (7.22). It follows that a particular solution of (7.19), in complex form, is given by

$$x_p = \frac{a-bi}{2} \frac{1}{p(i\omega)} e^{i\omega t} + \frac{a+bi}{2} \frac{1}{p(-i\omega)} e^{-i\omega t}$$

where $p(\lambda) = m\lambda^2 + r\lambda + k$.

We conclude with a particular case of (7.19) that requires special handling. Consider

$$\ddot{x} + x = \cos(t). \tag{7.23}$$

The form of the right-hand side suggests setting $x_p = \alpha \cos(t) + \beta \sin(t)$. However, that is the same as the complimentary function for (7.23), so it cannot be a particular solution, as is further evidenced by division by zero in (7.20). In order to solve (7.23), let us regard it as a limiting case of

$$\ddot{x} + x = \cos(\omega t). \tag{7.24}$$

As long as $\omega \neq 1$ the complimentary function of (7.24)

$$x_c = C_1 e^{it} + C_2 e^{-it} = a \cos(t) + b \sin(t)$$

is different from the particular solution (given by (7.20))

$$x_p = \frac{\cos(\omega t)}{1 - \omega^2}$$

and the solution of (7.24) with generic initial conditions $x(0) = x_0$ and $\dot{x}(0) = \dot{x}_0$ can be found without difficulty:

$$x = x_0 \cos(t) + \dot{x}_0 \sin(t) + \frac{\cos(\omega t) - \cos(t)}{1 - \omega^2}. \tag{7.25}$$

Taking the limit of (7.25) as $\omega \to 1$ gives the solution of (7.23) with the same generic initial conditions:

$$x = x_0 \cos(t) + \dot{x}_0 \sin(t) + \frac{t}{2} \sin(t). \tag{7.26}$$

The last term in Eq. (7.26) shows that a particular solution of (7.23) should be sought as the harmonic, suggested by the method of undetermined coefficients, but multiplied by t. Indeed, if we substitute $x_p = t(\alpha \cos(t) + \beta \sin(t))$ into (7.23) then, after equating like coefficients in $\ddot{x} + x = 2\beta \cos(t) + 2\alpha \sin(t) = \cos(t)$, we get $x_p = (t/2) \sin(t)$.

Equation (7.26) has interesting physical interpretation: in the absence of friction the displacement of the mass becomes unbounded if it is driven by a harmonic external force of the same frequency as the frequency of the complimentary function—the *natural frequency* of the mass-spring system. This is an instance of *resonance*.

7.5 Convolution in One Dimension

If the right-hand side of a nonhomogeneous ODE is too complicated for the method of undetermined coefficients, a particular solution can be computed as a convolution integral. We will first derive it for the first order scalar ODE

$$\frac{dx}{dt} + a x = f. \tag{7.27}$$

The derivation is inspired by the left endpoint rule of numerical integration whereby

$$\int_0^T f(t)\,dt \approx \sum_{k=0}^{N-1} f(k h)\, h, \quad h = \frac{T}{N}. \tag{7.28}$$

In Calculus (7.28) is narrowly associated with approximating integrals. However, the left endpoint rule is really a method of approximating functions, as we will now explain.

Let f be continuously differentiable on $[0, T]$—an assumption that is not strictly necessary but which will be useful in a later argument—and let $h = T/N$ be the grid size, as in (7.28). Define

$$f_h(t) = f(h \lfloor t/h \rfloor) \tag{7.29}$$

where $\lfloor \cdot \rfloor$ is the *floor function*—the function that outputs the greatest integer that is less than or equal to the input. The function f_h agrees with f at all but the last of the grid points $t_k = k h$ and maintains constant value $f(t_{k-1})$ on $[t_{k-1}, t_k)$. We can thus regard f_h as a piecewise constant approximation of f whose values are determined by the left endpoint rule.

Figure 7.3 shows the plots of (7.29) for $f(t) = \cos(3 t)$, $T = 2\pi$, and $h = 2^{-n}$ with $n = 0, 1, 2, 3$. The convergence is visually evident and can be quantified as follows.

Suppose that t lies in (t_{k-1}, t_k). According to the Mean Value Theorem (this is where we require f to be differentiable)

$$f(t) - f_h(t) = f(t) - f(t_{k-1}) = f'(\theta)\,(t - t_{k-1})$$

with θ somewhere between t_{k-1} and t. It follows that

$$|f(t) - f_h(t)| = |f'(\theta)|\,|t - t_{k-1}| \le \max_{t_{k-1} \le \theta \le t_k} |f'(\theta)|\, h.$$

and, consequently,

$$\max_{0 \le t \le T} |f(t) - f_h(t)| \le \max_k \max_{t_{k-1} \le \theta \le t_k} |f'(\theta)|\, h = \max_{0 \le \theta \le T} |f'(\theta)|\, h. \tag{7.30}$$

Now, since f is continuously differentiable, f' is continuous and hence attains maximum on $[0, T]$, by the Extreme Value Theorem. The maximum absolute error of approximating f

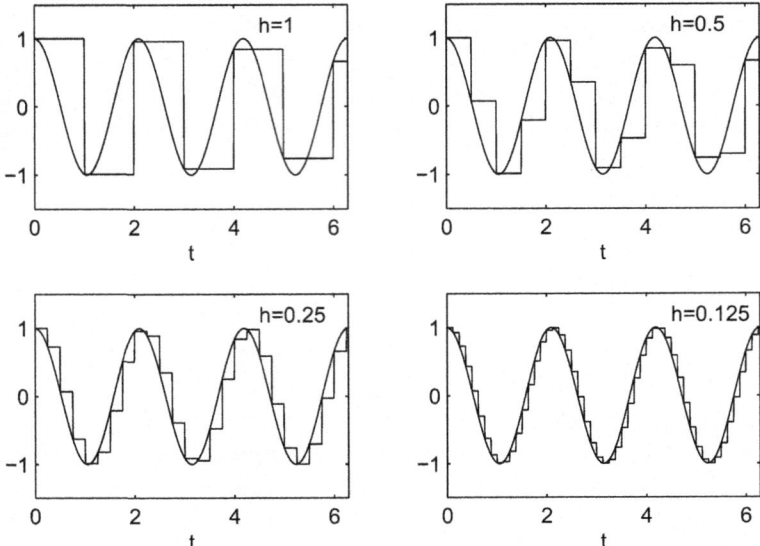

Fig. 7.3 Approximation of $f(t) = \cos(3t)$ on $[0, 2\pi]$ with step functions $f_h(t)$ defined by (7.29)

with f_h on $[0, T]$ is thus bounded by a constant—the maximum of the absolute value of the derivative—times the grid size. If the latter tends to zero, so does the error. This proves that $\lim_{h \to 0} f_h(t) = f(t)$.

Figure 7.4 is the log-log plot of $\max_{0 \leq t \leq T} |f(t) - f_h(t)|$ versus h for $f = \cos(3t)$ and $T = 2\pi$. As h decreases, the plot becomes a straight line, in accordance with (7.30). The linear fit computed using 10 circled points, corresponding to the 10 smallest values of h, shows that

$$\max_{0 \leq t \leq T} |f(t) - f_h(t)| \sim 2.9521\, h.$$

For $f = \cos(3t)$ the maximum absolute value of the derivative on $[0, 2\pi]$ is 3.

Using (7.29), we can restate the left endpoint rule of integration as

$$\int_0^T f(t)\, dt \approx \int_0^T f_h(t)\, dt.$$

In order to approximate the integral of a function, we approximate the function with a step function which is simple to integrate and whose integral is, in fact, the sum in (7.28). Similar logic can be applied to ODE (7.27). We can derive its particular solution by solving the simpler ODE

$$\frac{dx_h}{dt} + a\, x_h = f_h \qquad (7.31)$$

and taking the limit of x_h as $h \to 0$, as illustrated in Fig. 7.5.

7.5 Convolution in One Dimension

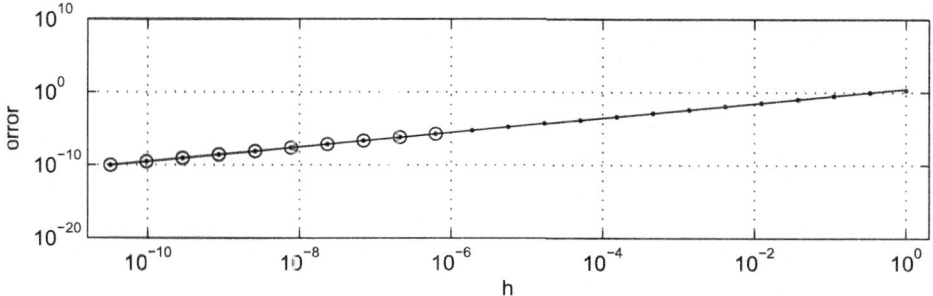

Fig. 7.4 Log-log plot of the error $\max_{0 \leq t \leq T} |f(t) - f_h(t)|$ against the grid size h for $f = \cos(3t)$ and $T = 2\pi$; also shown is a linear fit computed using the circled data points

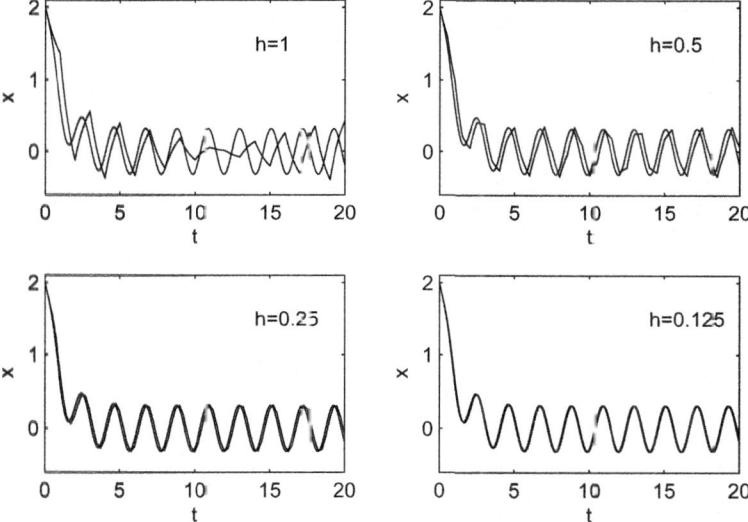

Fig. 7.5 Solutions of (7.27) and (7.31) with $f(t) = \cos(3t)$, $a = 1$, $x_0 = 2$

Figure 7.5 shows the solutions of (7.27) and (7.31) for $f(t) = \cos(3t)$, $a = 1$, $x_0 = 2$, and $h = 2^{-n}$ with $n = 0, 1, 2, 3$. It is evident that, as $h \to 0$, the solution of (7.31) (computed using the formula which we are about to derive) approaches the solution of (7.27) (whose formula we derived in Sect. 6.7 through the first use of the method of undetermined coefficients).

Let us construct a particular solution of ODE (7.31) corresponding to zero initial condition. Since the right-hand side of (7.31) is constant on intervals $[t_{k-1}, t_k)$, that can be done interval-by-interval.

On the first interval $t_0 = 0 \leq t < t_1$ the right-hand side of (7.31) is $f(0)$, and the solution with zero initial condition is

$$x_{h,1} = \frac{f(0)}{a}\left(1 - e^{-at}\right), \quad 0 \le t < t_1.$$

On the second interval $t_1 \le t < t_2$ the right-hand side of (7.31) is $f(t_1)$. Since x_h is continuous, the initial condition at $t = t_1$ must agree with the value of the solution on the first interval: $x_{h,2}(t_1) = x_{h,1}(t_1) = \frac{f(0)}{a}\left(1 - e^{-at_1}\right)$. In order to reveal the pattern, we will write

$$x_{h,2} = \frac{f(0)}{a}\left(e^{-a(t-t_1)} - e^{-at}\right) + \frac{f(t_1)}{a}\left(1 - e^{-a(t-t_1)}\right), \quad t_1 \le t < t_2.$$

On the third interval $t_2 \le t < t_3$ the right-hand side is $f(t_2)$, the initial condition is $x_{h,3}(t_2) = x_{h,2}(t_2)$, and the solution is

$$\begin{aligned}x_{h,3} = &\frac{f(0)}{a}\left(e^{-a(t-t_1)} - e^{-at}\right) + \frac{f(t_1)}{a}\left(e^{-a(t-t_2)} - e^{-a(t-t_1)}\right) \\ &+ \frac{f(t_2)}{a}\left(1 - e^{-a(t-t_2)}\right), \quad t_2 \le t < t_3.\end{aligned}$$

For the forth interval,

$$\begin{aligned}x_{h,4} = &\frac{f(0)}{a}\left(e^{-a(t-t_1)} - e^{-at}\right) + \frac{f(t_1)}{a}\left(e^{-a(t-t_2)} - e^{-a(t-t_1)}\right) \\ &+ \frac{f(t_2)}{a}\left(e^{-a(t-t_3)} - e^{-a(t-t_2)}\right) + \frac{f(t_3)}{a}\left(1 - e^{-a(t-t_3)}\right), \quad t_3 \le t < t_4.\end{aligned}$$

and so on. The pattern is now apparent

$$\begin{aligned}x_{h,k} = &\sum_{l=0}^{k-2} \frac{f(t_l)}{a}\left(e^{-a(t-t_{l+1})} - e^{-a(t-t_l)}\right) \\ &+ \frac{f(t_{k-1})}{a}\left(1 - e^{-a(t-t_{k-1})}\right), \quad t_{k-1} \le t < t_k,\end{aligned} \quad (7.32)$$

and can be proved by induction.

A less straightforward, but more elegant way of solving (7.31) is to use the principle of superposition; we will present it because it reinforces core concepts of Chap. 6 and gives the solution in a form that is easier to analyze.

Let $\phi_k(t) = \phi(t - t_k)$ where ϕ is the *characteristic function* of the interval $[0, t_1)$, which is to say it is 1 on $[0, t_1)$ and zero everywhere else:

$$\phi(t) = \begin{cases} 1, & \text{if } 0 \le t < t_1, \\ 0, & \text{otherwise.} \end{cases}$$

The function ϕ_k is 1 on $[t_k, t_{k+1})$ and zero everywhere else: it is the characteristic function of $[t_k, t_{k+1})$. The left endpoint approximation f_h is a linear combination of ϕ_k's: $f_h = \sum_{k=0}^{N-1} f(t_k)\phi_k$. According to the principle of superposition, the solution of (7.31) (with zero initial condition) is

7.5 Convolution in One Dimension

$$x_h(t) = \sum_{k=0}^{N-1} f(t_k)\,\xi_k(t), \tag{7.33}$$

where ξ_k is the solution of

$$\frac{d\xi_k}{dt} + a\,\xi_k = \phi_k, \quad \xi(0) = 0,$$

which is given by

$$\xi_k(t) = \frac{1}{a} \begin{cases} 0, & 0 \le t < t_k, \\ 1 - e^{-a(t-t_k)}, & t_k \le t < t_{k+1}, \\ e^{-a(t-t_{k+1})} - e^{-a(t-t_k)}, & t_{k+1} \le t < T, \end{cases} \tag{7.34}$$

and has the profile shown in Fig. 7.6.

Using (7.33), we can express the particular solution of (7.27) as the limit

$$x(t) = \lim_{h \to 0} x_h(t) = \lim_{N \to \infty} \sum_{k=0}^{N-1} f(t_k)\,\xi_k(t) \tag{7.35}$$

whose semblance to a limit of a Riemann sum strongly suggests that it should be an integral involving f.

Let us evaluate the limit (7.35) at $t = T$. According to (7.34), and as can be seen from Fig. 7.6,

$$\xi_k(T) = \frac{1}{a}\left(e^{-a(T-t_{k+1})} - e^{-a(T-t_k)}\right), \quad k = 0, \ldots, N-1.$$

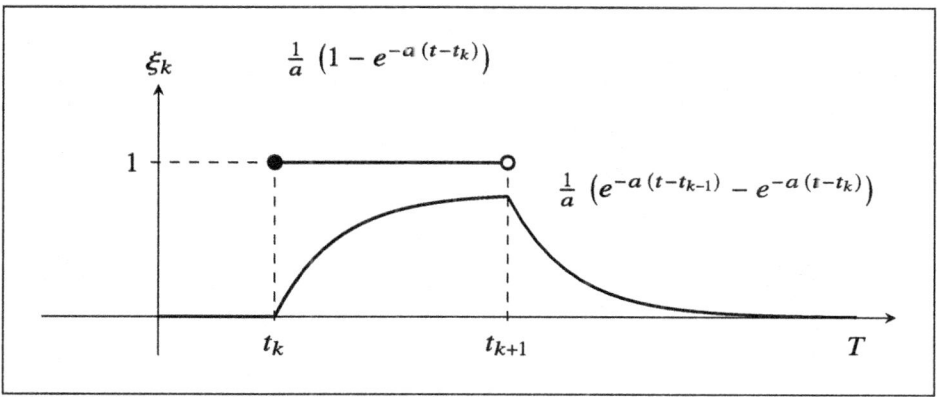

Fig. 7.6 Typical plot of (7.34)

Hence,

$$x_h(T) = \frac{1}{a} \sum_{k=0}^{N-1} f(t_k) \left(e^{-a(T-t_{k+1})} - e^{-a(T-t_k)} \right)$$ (7.36)

$$= e^{-aT} \sum_{k=0}^{N-1} f(t_k) e^{a t_k} \frac{e^{a h} - 1}{a h} h.$$

Since

$$\lim_{h \to 0} \frac{e^{a h} - 1}{a h} = 1$$

it follows that

$$\lim_{N \to \infty} \sum_{k=0}^{N-1} f(t_k) e^{a t_k} \frac{e^{a h} - 1}{a h} h = \lim_{N \to \infty} \sum_{k=0}^{N-1} f(t_k) e^{a t_k} h = \int_0^T f(t) e^{a t} dt.$$

Consequently,

$$x(T) = e^{-aT} \int_0^T f(t) e^{a t} dt = \int_0^T f(t) e^{-a(T-t)} dt.$$ (7.37)

Equation (7.37) was derived under the assumption that T is constant, however, since T is set arbitrarily, we can replace it with a variable t; the variable of integration in (7.37) should then be changed to some other variable, say s. We conclude that the particular solution of (7.27) corresponding to zero initial condition is

$$x(t) = \int_0^t f(s) e^{-a(t-s)} ds.$$ (7.38)

The integral in (7.38) defines a binary operation on functions called *convolution*. In general, the convolution of two functions f and g is a third function $f * g$ defined as the integral

$$f * g = \int_0^t f(s) g(t-s) ds.$$ (7.39)

Equation (7.38) is a special case of (7.39); in $*$-notation it reads: $x = f * e^{-at}$.

In principle, we found a general formula for a particular solution of (7.27) satisfying zero initial condition. However, the main utility of Eq. (7.38) does not lie, as one might think, in finding particular solutions of (7.27). To see that, consider the following instance of (7.27) where the right-hand side is a simple power:

$$\frac{dx}{dt} + x = t^3$$

We can immediately write the general solution as

7.5 Convolution in One Dimension

$$x = Ce^{-t} + e^{-t} * t^3 = Ce^{-t} + \int_0^t e^{-(t-s)} s^3 \, ds,$$

but computing the convolution integral requires triple integration by parts. It is much more expedient to guess $x_p = c_0 + c_1 t + c_2 t^2 + c_3 t^3$ and find the coefficients by solving the linear system

$$c_3 = 1, \quad 3c_3 + c_2 = 2c_2 + c_1 = c_1 + c_0 = 0$$

which immediately gives $x_p = -6 + 6t - 3t^2 + t^3$. Whenever the right-hand side of (7.27) is elementary, it is faster to use the method of undetermined coefficients than to integrate.

Next, consider this instance of (7.27) with non-elementary right-hand side:

$$\frac{dx}{dt} + x = e^{-t^3}.$$

Again, we can write

$$x = Ce^{-t} + e^{-t} * e^{-t^3} = Ce^{-t} + \int_0^t e^{-(t-s)} e^{-s^3} \, ds,$$

but the integral is not expressible in closed form and has to be computed numerically; unless a very precise computation is required, it is simpler to numerically solve the ODE.

While Eq. (7.38) may not be the best way to compute particular solutions for concrete instances of ODE (7.27), it does establish an important relation between the solution x of the IVP

$$\frac{dx}{dt} + ax = f, \quad x(0) = 0,$$

and the right-hand side f; the function f is often regarded as the "input" while x is the "output" of some physical system. With the help of (7.38) we can make constructive statements about x based on limited information about f. For instance, suppose that all we know about f is that its magnitude is bounded by some constant: $|f| \leq C$. What can we say about the particular solution of (7.27) based on that?

As follows from (7.38),

$$|x| = \left| \int_0^t f(s) e^{-a(t-s)} \, ds \right| = e^{-at} \left| \int_0^t f(s) e^{as} \, ds \right|$$

$$\leq e^{-at} \int_0^t |f(s)| e^{as} \, ds \leq e^{-at} \int_0^t C e^{as} \, ds = C \frac{1 - e^{-at}}{a}.$$

If $a > 0$, in which case (7.27) is said to be *stable*, then we can further deduce that $|x| \leq C/a$. Thus we can relate the maximum output of the system to the maximum value of the input without knowing all of the details of the latter.

7.6 Multidimensional Convolution

The convolution integral for the solution of the matrix-vector IVP

$$\frac{d\mathbf{u}}{dt} = A\mathbf{u} + \mathbf{g}, \quad \mathbf{u}(0) = \mathbf{0} \tag{7.40}$$

is similar to (7.38). It can be derived using the same technique as in Sect. 7.5, but we will present a simpler, albeit nonconstructive derivation.

Let us seek the solution of (7.40) in the form $\mathbf{u} = e^{tA} \mathbf{c}(t)$ where $\mathbf{c}(t)$ is a function to be determined. Notice that we replaced the constant $\mathbf{c}$ in the complimentary function $\mathbf{u}_c = e^{-tA} \mathbf{c}$ with a function of time—this is a technique known as *Variation of Parameters*. Substituting this expression into (7.40) results in

$$A e^{tA} \mathbf{c}(t) + e^{tA} \frac{d\mathbf{c}(t)}{dt} = A e^{tA} \mathbf{c}(t) + \mathbf{g},$$

which upon simplification becomes

$$\frac{d\mathbf{c}}{dt} = e^{-tA} \mathbf{g}.$$

It follows that

$$\mathbf{c}(t) = \int_0^t e^{-sA} \mathbf{f}(s) \, ds,$$

(the zero lower limit ensures that $\mathbf{u}(0) = \mathbf{0}$) and therefore the solution of (7.40) is

$$\mathbf{u}(t) = \int_0^t e^{(t-s)A} \mathbf{g}(s) \, ds. \tag{7.41}$$

Equation (7.41) looks like (7.38), however, it should be written with care: the matrix exponential $e^{(t-s)A}$ must precede the vector $\mathbf{f}(s)$ inside the integral because of the row-by-column rule of matrix multiplication.

Just as its one-dimensional counterpart (7.38), multidimensional convolution (7.41) is not particularly useful for solving ODE for the same reasons that are listed at the end of Sect. 7.5. It does, however, establish an important relation between the forcing term $\mathbf{g}$ in (7.40) and the solution $\mathbf{u}$ which, as we have already mentioned, are often regarded as input and output of some system.

As an example of using (7.41), consider the mass-spring system in Fig. 7.1. Suppose that it is disturbed from rest by an external force f and that we require the resulting displacement x. Newton's second law

$$m\ddot{x} + r\dot{x} + kx = f, \quad x(0) = \dot{x}(0) = 0,$$

when written in matrix-vector form

7.7 Impulse Response

$$\frac{d}{dt}\begin{bmatrix} x \\ \dot{x} \end{bmatrix} = \begin{bmatrix} 0 & 1 \\ -\frac{k}{m} & -\frac{r}{m} \end{bmatrix} \begin{bmatrix} x \\ \dot{x} \end{bmatrix} + \begin{bmatrix} 0 \\ \frac{1}{m}f \end{bmatrix}, \quad \begin{bmatrix} x(0) \\ \dot{x}(0) \end{bmatrix} = \begin{bmatrix} 0 \\ 0 \end{bmatrix},$$

is an instance of (7.40) with

$$\mathbf{u} = \begin{bmatrix} x \\ \dot{x} \end{bmatrix}, \quad A = \begin{bmatrix} 0 & 1 \\ -\frac{k}{m} & -\frac{r}{m} \end{bmatrix}, \quad \mathbf{g} = \begin{bmatrix} 0 \\ \frac{1}{m}f \end{bmatrix}.$$

Hence, according to (7.41),

$$\begin{bmatrix} x(t) \\ \dot{x}(t) \end{bmatrix} = \int_0^t \exp\left((t-s)\begin{bmatrix} 0 & 1 \\ -\frac{k}{m} & -\frac{r}{m} \end{bmatrix}\right) \begin{bmatrix} 0 \\ \frac{1}{m}f(s) \end{bmatrix} ds. \quad (7.42)$$

In Sect. 7.1, where the mass-spring system was first introduced, we found the eigendecomposition of the matrix A (Eq. (7.12)) and used it to compute e^{tA} (Eq. (7.13)). Replacing t with $t-s$ in (7.13) gives the matrix $e^{(t-s)A}$ inside the integral in (7.42). Carrying out matrix-vector multiplication and extracting the first component corresponding to displacement yields

$$x(t) = \int_0^t \frac{1}{m} \frac{e^{\lambda_2 (t-s)} - e^{\lambda_1 (t-s)}}{\lambda_2 - \lambda_1} f(s) \, ds, \quad (7.43)$$

where $\lambda_{1,2}$ are the eigenvalues given by (7.11).

7.7 Impulse Response

Equation (7.43) has the form $x(t) = \int_0^t G(t-s) f(s) \, ds$ where

$$G(t) = \frac{1}{m} \frac{\lambda_2 e^{\lambda_1 t} - \lambda_1 e^{\lambda_2 t}}{\lambda_2 - \lambda_1} \quad (7.44)$$

is the *impulse response* of the mass-spring system. If the force f is applied to the cart for an infinitesimally brief time, but in such a manner that the resulting change in the cart's momentum is unity, Eq. (7.44) gives the displacement of the cart. To see that, suppose that the force is

$$f(t) = \frac{1}{\tau}\begin{cases} 1, & \text{if } 0 \le t \le \tau, \\ 0, & \text{otherwise}. \end{cases} \quad (7.45)$$

Then, according to (7.43), the displacement is

$$x_\tau(t) = \frac{1}{m(\lambda_2 - \lambda_1)}\left\{\frac{\lambda_2}{\lambda_1}\frac{e^{\lambda_1 (t-\tau)} - e^{\lambda_1 t}}{\tau} - \frac{\lambda_1}{\lambda_2}\frac{e^{\lambda_2 (t-\tau)} - e^{\lambda_2 t}}{\tau}\right\}.$$

Taking the limit of this expression as $\tau \to 0$ gives (7.44): $\lim_{\tau \to 0} x_\tau(t) = G(t)$.

The significance of the impulse response (7.44) is that once it is known the response of the cart to any force can be computed using (7.43). However, measuring impulse response directly, using an input like (7.45) can be challenging. In practice, impulse response is often computed as the derivative of the *step response*—the response to the *unit step function*:

$$u(t) = \begin{cases} 1, & \text{if } t \geq 0, \\ 0, & \text{otherwise}. \end{cases} \tag{7.46}$$

The computation is based on the identity $\frac{d}{dt}\int_0^t G(t-s)\,ds = G(t)$. While mass-spring systems are easier to visualize, performing measurements is easier for circuits. We will therefore illustrate the step and impulse response using an RLC-circuit.

7.8 Step and Impulse Response of an RLC-Circuit

Figure 7.7 shows the voltage across the capacitor in an RLC-circuit driven by a square waveform as well as the square waveform itself.

The elements of the circit, connected as shown in Fig. 7.2, have nominal parameters R = 100 Ω, L = 1 mH, and C = 470 pF; the frequency of the square waveform is 10 kHz.

As the square waveform jumps from 0V to 1V the circuit responds by "ringing". There are two such transitions in Fig. 7.7 and either one can be used as an approximation of the step response; we used the first transition which is shown separately in Fig. 7.8.

Theoretically, the step response, which is shown in Fig. 7.8, is modeled by the IVP

$$LC\frac{d^2V}{dt^2} + RC\frac{dV}{dt} + V = 1, \quad V(0) = \frac{dV}{dt}(0) = 0. \tag{7.47}$$

Equation (7.47) follows from (7.17) after one sets $V = Q/C$ and adds external unit voltage on the right-hand side. The solution of (7.47), which can be quickly derived using the shortcut explained in Sect. 7.3 and the method of undetermined coefficients, is given by

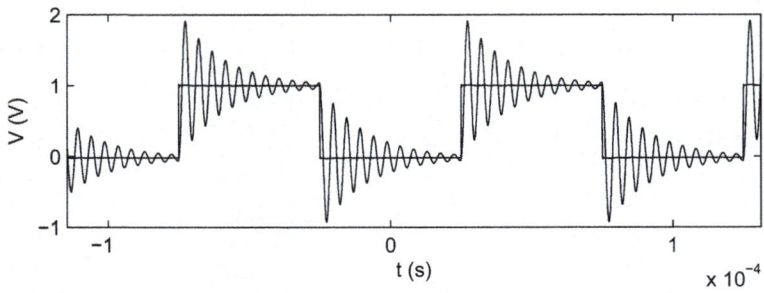

Fig. 7.7 Response of an RLC-circuit to a square wave

7.8 Step and Impulse Response of an RLC-Circuit

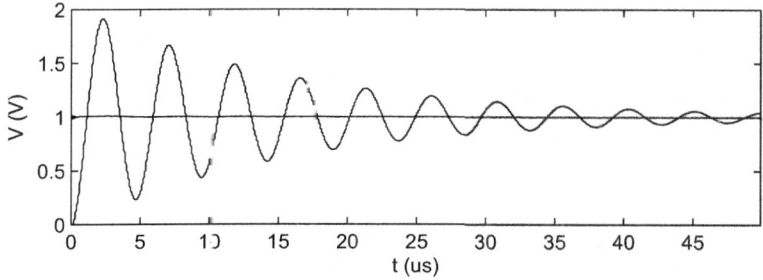

Fig. 7.8 Step response extracted from the data in Fig. 7.7

$$V = 1 + \frac{1}{\lambda_1 - \lambda_2} \left(\lambda_2 e^{\lambda_1 t} - \lambda_1 e^{\lambda_2 t} \right), \quad (7.48)$$

with $\lambda_{1,2}$ being the roots of $LC\lambda^2 + RC\lambda + V = 0$. The characteristic roots are complex conjugates $\lambda_{1,2} = a \pm bi$ and (7.48) may be rewritten in a more revealing way as

$$V = 1 + e^{at} \left(-\cos(bt) + a \frac{\sin(bt)}{b} \right). \quad (7.49)$$

If we use nominal values of the circuit elements then

$$a = -\frac{R}{2L} = 5 \times 10^4 \, 1/s, \quad b = \sqrt{\frac{1}{LC} - \left(\frac{R}{2L}\right)^2} \approx 1.4578 \times 10^6 \, 1/s. \quad (7.50)$$

Of course, none of the values are likely to be nominal and therefore the actual values of a and b must be determined through a two-parameter nonlinear fit.

We refined the starting values given by (7.50) using fminsearch before supplying them to Newton's method which then converged to

$$a = -6.1770660495164 \times 10^4 \, 1/s, \quad b = 1.327882962152141 \times 10^6 \, 1/s$$

6 after 3 iterations. The resulting fit and the residual are shown in Fig. 7.9. Since the residual is far from being white noise, its statistical analysis does not give useful information and is omitted.

Having estimated the characteristic roots, we can now compute the impulse response as the derivative of the step response:

$$G(t) = \frac{dV}{dt} = \frac{\lambda_1 \lambda_2}{\lambda_1 - \lambda_2} \left(e^{\lambda_1 t} - e^{\lambda_2 t} \right) = \left(a^2 + b^2 \right) e^{at} \frac{\sin(bt)}{b}. \quad (7.51)$$

The plot of (7.51) is shown in Fig. 7.10. Notice the y-scale: while the step response is measured in volts, the impulse response is measured in megavolts-per-second. Measuring values at such scale directly is very challenging, even with sophisticated instrumentation.

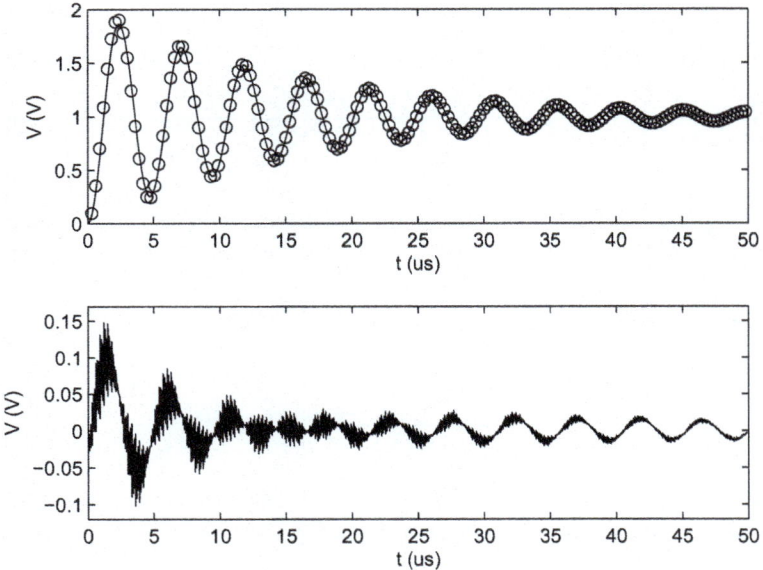

Fig. 7.9 The top panel shows (7.49) (solid line) fitted to the data in Fig. 7.8 (circles); to avoid clutter, we plotted every 10-th data point. The bottom panel shows the residual

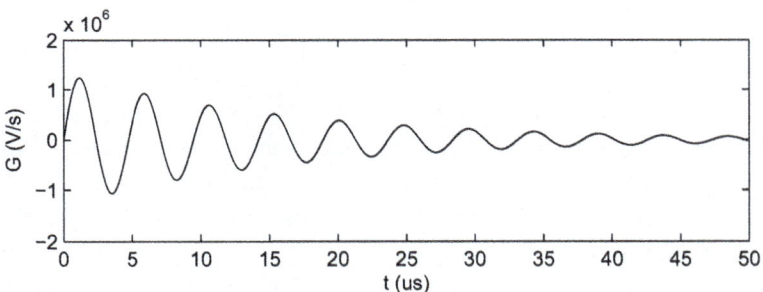

Fig. 7.10 Impulse response (7.51) computed from the step response

When the square waveform transitions from low voltage to high voltage, neither the voltage across the capacitor V nor its derivative $\dot{V}$ are exactly zero, as we assume in (7.47); nor are the low and high voltages of the square waveform exactly 0V and 1V, respectively. One way to take that into account is to replace (7.49) with a more general formula

$$V = c + e^{at}\left((V_0 - c)\cos(bt) + \left(\dot{V}_0 - a(V_0 - c)\right)\frac{\sin(bt)}{b}\right).$$

Alternatively, we can decrease the frequency of the square wave. This will allow the circuit to dissipate more energy during the low voltage part of the square waveform cycle making zero initial conditions in (7.47) more realistic. Unfortunately, the potential improvements are hardly worth the effort because of other sources of error. In addition to probe loading, nonlinearities of circuit elements, and other factors listed in Sect. 3.3, we have to contend with the square waveform not being perfectly square: what we see in Fig. 7.7 is actually a smooth approximation to a discontinuous voltage that cannot be physically generated.

Based on the fit in the top panel of Fig. 7.9, we do have certainty that

$$a \approx -6.2 \times 10^4 \text{ 1/s}, \quad b \approx 1.3 \times 10^6 \text{ 1/s}.$$

However the impulse response shown in Fig. 7.10 is unlikely to be accurate enough for practical computations.

7.9 Comments and Bibliography

According to [5], the first complete treatment of $d\mathbf{x}/dt = A\mathbf{x}$ was given by Weierstrass in the second volume of his "Werke" in 1875. However, the idea of eigendecomposition goes back to Cauchy's 1829 paper [1] which dealt with secular (non-periodic) perturbations of planetary orbits. Actually, Cauchy's main interest was in finding principle axes of inertia tensors. To that end, he proved (in modern language) that symmetric matrices have real eigenvalues and mutually orthogonal eigenvectors.

Eigendecomposition is not limited to matrices, but is applicable to all linear operators. Let

$$L : x \mapsto a_n x^{(n)} + a_{n-1} x^{(n-1)} + a_{n-2} x^{(n-2)} + \cdots + a_0 x$$

be a linear operator with constant coefficients. Since

$$L\left(e^{\lambda t}\right) = \left(a_n \lambda^n + a_{n-1} \lambda^{n-1} + a_{n-2} \lambda^{n-2} + \cdots + a_0\right) e^{\lambda t} = p(\lambda) e^{\lambda t}$$

the exponential function $e^{\lambda t}$ is an *eigenfunction* of L with eigenvalue $p(\lambda)$ where p is the characteristic polynomial. If $p(\lambda) = 0$ then $L(e^{\lambda t}) = 0$ and $e^{\lambda t}$ is a solution of the homogeneous ODE $L(x) = 0$: this is the linear algebra perspective on the shortcut to general solutions of homogeneous ODE discussed in Sect. 7.3. By the way, the words "eigenvalue" and "eigenfunction" were inspired by Hilbert's terminology in [4] which was published in 1904 (the word "eigenvector" appeared after "eigenfunction"). Prior to the work of Hilbert, in place of the German prefix "eigen" (which roughly translates as "self") mathematicians used "secular" (after the astronomy researches of Cauchy), "latent", "proper", "characteristic", and "singular." After the Second World War there was a movement among British and American mathematicians to remove German words from the mathematical vocabulary. Luckily, common sense prevailed and the distinct and unambiguous eigen-terminology retained its rightful place in linear algebra.

We should also mention that the idea of eigendecomposition is closely related (and historically preceded) by diagonalization of quadratic forms. Suppose that $f(x, y)$ has a critical point which, for simplicity, can be located at the origin $(0, 0)$. In the vicinity of the critical point the shape of the graph of $f(x, y)$ is that of the Taylor quadratic

$$\frac{1}{2}\frac{\partial^2 f}{\partial x^2}(0, 0)\, x^2 + \frac{\partial^2 f}{\partial x \partial y}(0, 0)\, x\, y + \frac{1}{2}\frac{\partial^2 f}{\partial y^2}(0, 0)\, y^2 = \frac{1}{2}\left(a\, x^2 + 2\, b\, x\, y + c\, y^2\right).$$

Rewriting the Taylor quadratic as

$$\frac{1}{2}\begin{bmatrix} x & y \end{bmatrix}^T \begin{bmatrix} a & b \\ b & c \end{bmatrix} \begin{bmatrix} x \\ y \end{bmatrix}$$

reduces the classification of the critical point to the analysis of the Hessian matrix. Since the Hessian is symmetric, it is diagonalizable, with real eigenvalues; furthermore, it is diagonalized by an *orthogonal matrix* whose inverse is its transpose:

$$\begin{bmatrix} a & b \\ b & c \end{bmatrix} = P \begin{bmatrix} \lambda_1 & 0 \\ 0 & \lambda_2 \end{bmatrix} P^T.$$

Consequently, we can rewrite the Taylor quadratic as

$$\frac{1}{2}\begin{bmatrix} x & y \end{bmatrix}^T P \begin{bmatrix} \lambda_1 & 0 \\ 0 & \lambda_2 \end{bmatrix} P^T \begin{bmatrix} x \\ y \end{bmatrix} = \frac{1}{2}\begin{bmatrix} u & v \end{bmatrix}^T \begin{bmatrix} \lambda_1 & 0 \\ 0 & \lambda_2 \end{bmatrix} \begin{bmatrix} u \\ v \end{bmatrix} = \lambda_1 u^2 + \lambda_2 v^2,$$

where (u, v) are the new coordinates related to the (x, y)-coordinates by means of the matrix P^T. If both eigenvalues of the Hessian are positive, then the shape of $\lambda_1 u^2 + \lambda_2 v^2$ is that of an upward-opening elliptic paraboloid and the critical point is a local minimum; if both eigenvalues are negative the paraboloid opens downward and the critical point is a local maximum; finally, if the eigenvalues have opposite signs, the quadratic surface is a hyperbolic paraboloid and the critical point is a saddle point. Classification of higher-dimensional critical points is just as simple: one computes the eigenvalues of the Hessian, which were proven to be real by Cauchy, and checks their signs.

In order to define impulse response in Sect. 7.7, we modeled impulsive force as the limit of (7.45). This kind of limit was considered by Dirac in his seminal "The Principles of Quantum Mechanics" [2] and is called, misleadingly, the *Dirac delta function*:

$$\delta(t) = \lim_{\tau \to 0} \frac{1}{\tau} \begin{cases} 1, & \text{if } 0 \leq t \leq \tau, \\ 0, & \text{otherwise.} \end{cases} \tag{7.52}$$

The misleading word is "function": the limit (7.52) is not a function in the classical sense but is a more complicated object called a *distribution*.

Dirac thought of (7.52) as a special kind of function defined on the entire real line with the property

$$\int_{-\infty}^{\infty} \delta(t)\, f(t)\, dt = f(0) \tag{7.53}$$

holding for every "regular" function $f(t)$. As follows from (7.53),

$$\int_{-\infty}^{\infty} \delta(t-s) f(s) \, ds = \int_{-\infty}^{\infty} \delta(s) f(t-s) \, ds = f(t).$$

Consequently, if L is a linear differential operator with constant coefficients and $L(G) = \delta(t)$ then

$$x(t) = \int_{-\infty}^{\infty} G(t-s) f(s) \, ds$$

is a solution of $L(x) = f$. Indeed,

$$L\left(\int_{-\infty}^{\infty} G(t-s) f(s) \, ds\right) = \int_{-\infty}^{\infty} L(G(t-s)) f(s) \, ds = \int_{-\infty}^{\infty} \delta(t-s) f(s) \, ds = f(t).$$

According to Dirac, impulse response for a system modeled by $L(x) = f$ is the solution of $L(G) = \delta(t)$ satisfying certain conditions imposed by physics, such as *causality*. This is an elegant definition which is often used to derive formulas such as (7.38). In our derivation of (7.38) in Sect. 7.5 we used only simple Calculus because proper understanding of δ-functions requires extensive foray into mathematical analysis. We will similarly avoid distributions in the rest of the book but an interested reader will find the relevant theory in [3] one of whose authors, Laurent Schwartz, won the Fields Medal for putting Dirac's largely intuitive δ-calculus on firm mathematical footing.

7.10 Exercises

1. How fast does the solution x_h approach x in Fig. 7.5? Formulate a conjecture, after a few numerical experiments, and then try to prove it.
2. Derive the convolution integral (7.38) using piecewise linear rather than piecewise constant approximation. That is, form an approximating function f_h by linearly interpolating the values of f on equispaced grid.
3. Show that one dimensional convolution is commutative. *Hint:* Transform the integral $f * g = \int_0^t f(s) g(t-s) \, ds$ using a substitution.
4. Show that one dimensional convolution is associative. *Hint:* Convert $f * g * h$ into a double integral and interchange the order of integration.
5. Let $\chi(t)$ be the characteristic function of the unit interval

$$\chi(t) = \begin{cases} 1, & \text{if } 0 \leq t \leq 1 \\ 0, & \text{otherwise.} \end{cases}$$

Set $B_0 = \chi$ and define recursively $B_{n+1} = \chi * B_n$. Plot B_1, B_2 and B_3: these are called *B-splines*.
6. Find the step and impulse responses of an RC-circuit.

7 Solve
$$x^{(4)} + x = f, \quad x(0) = x^{(1)}(0) = x^{(2)}(0) = x^{(3)}(0) = 0$$
using convolution. Test the answer $x(t) = \int_0^t G(t-s) f(s) \, ds$ by setting f to a simple harmonic and comparing the integral with the solution derived using the method of undetermined coefficients.

8 Show that the solution of
$$\frac{d^2}{dt^2} \begin{bmatrix} x \\ y \end{bmatrix} = \begin{bmatrix} -2 & 1 \\ 1 & -2 \end{bmatrix} \begin{bmatrix} x \\ y \end{bmatrix}, \quad \begin{bmatrix} x(0) \\ y(0) \end{bmatrix} = \begin{bmatrix} x_0 \\ y_0 \end{bmatrix}, \quad \begin{bmatrix} \frac{dx}{dt}(0) \\ \frac{dy}{dt}(0) \end{bmatrix} = \begin{bmatrix} 0 \\ 0 \end{bmatrix}$$
is given by
$$\begin{bmatrix} x \\ y \end{bmatrix} = \cos\left(t \begin{bmatrix} -2 & 1 \\ 1 & -2 \end{bmatrix}\right) \begin{bmatrix} x_0 \\ y_0 \end{bmatrix}.$$
Find an explicit formula for $x(t)$ and confirm it in MATLAB.

References

1. A.-L. Cauchy, *Sur l'équation à l'aide de laquelle on détermine les inégalités séculaires des mouvements des planètes*. Cambridge Library Collection - Mathematics, vol. 9, pp. 174–195 (Cambridge University Press, 2009)
2. P.A.M. Dirac, *The Principles of Quantum Mechanics* (Clarendon Press, Oxford, 1930)
3. I. Halperin, L. Schwartz, *Introduction to the Theory of Distributions* (University of Toronto Press, 1952)
4. D. Hilbert, *Grundzüge einer allgemeinen Theorie der linearen Integralgleichungen*, pp. 8–171 (Vieweg+Teubner Verlag, Wiesbaden, 1989)
5. J.A. Nyswander, A direct solution of systems of linear differential equations having constant coefficients. Amer. J. Math. **47**(4), 257–276 (1925)

Discrete Fourier Transform 8

The chapter begins with a discussion of orthogonal expansions in $\mathbb{R}^n$ in Sect. 8.1. Discrete Fourier transform (DFT) may be regarded as one such expansion, however, as we explain in Sect. 8.2, it is better to think of it as a method for approximating functions with sums of harmonics. Since real harmonics can be expressed in terms of complex exponentials, real DFT can be reformulated as an orthogonal expansion in $\mathbb{C}^n$. In fact, most computer implementation of DFT use complex arithmetic.

The importance of DFT is hard to overstate. For us it will be mainly a stepping stone to Fourier series. Yet, it is also one of the most frequently used algorithms in data analysis. This is illustrated in Sect. 8.2 with examples showing how DFT is used to filter out noise, compare frequency contents of different data sets, and forecast periodic events. We also show how to approximate a solution of an ODE using DFT which foreshadows the developments in Chaps. 9 and 10.

8.1 Orthogonal Expansions in $\mathbb{R}^N$

Let $\{\mathbf{v}_i\}_{i=1}^N$ be a collection of N mutually orthogonal vectors in $\mathbb{R}^N$:

$$\mathbf{v}_i \cdot \mathbf{v}_j = 0, \quad i \neq j.$$

If none of the vectors are zero, as we assume, then, due to mutual orthogonality, they are linearly independent. Indeed, suppose that $\sum_{i=1}^N c_i \mathbf{v}_i = 0$. Then

$$\sum_{i=1}^N c_i \mathbf{v}_i \cdot \sum_{i=1}^N c_i \mathbf{v}_i = \sum_{i=1}^N c_i^2 \|\mathbf{v}_i\|^2 = 0$$

© The Author(s), under exclusive license to Springer Nature Switzerland AG 2025
A. Beltukov, *Differential Equations and Data Analysis*, Synthesis Lectures on Mathematics & Statistics, https://doi.org/10.1007/978-3-031-62257-1_8

implies that $c_i = 0$ for $i = 1, \ldots, N$.

Being linearly independent, the vectors $\{\mathbf{v}_i\}_{i=1}^{N}$ form a basis of $\mathbb{R}^N$. Any vector $\mathbf{w}$ in $\mathbb{R}^N$ is therefore expressible as a linear combination

$$\mathbf{w} = \sum_{i=1}^{N} c_i \mathbf{v}_i. \tag{8.1}$$

Dotting both sides of (8.1) with $\mathbf{v}_j$ and using mutual orthogonality of $\mathbf{v}_i$'s gives

$$\mathbf{v}_j \cdot \mathbf{w} = \mathbf{v}_j \cdot \sum_{i=1}^{N} c_i \mathbf{v}_i = \sum_{i=1}^{N} c_i \mathbf{v}_i \cdot \mathbf{v}_j = c_j \mathbf{v}_j \cdot \mathbf{v}_j$$

whence follows that $c_j = \mathbf{v}_j \cdot \mathbf{w} / \mathbf{v}_j \cdot \mathbf{v}_j$ and hence

$$\mathbf{w} = \sum_{i=1}^{N} \frac{\mathbf{v}_i \cdot \mathbf{w}}{\mathbf{v}_i \cdot \mathbf{v}_i} \mathbf{v}_i. \tag{8.2}$$

Equation (8.2) is a general formula for orthogonal expansion in $\mathbb{R}^N$. The terms in the sum are orthogonal projections of $\mathbf{w}$ onto $\mathbf{v}_i$'s.

A basis whose vectors are mutually orthogonal is called, naturally, an *orthogonal basis*. The standard basis of $\mathbb{R}^n$ is an orthogonal basis. It is used almost exclusively in Calculus and Physics, however, apart from its simplicity, there is nothing exceptional about it— any collection of N mutually orthogonal vectors forms an orthogonal basis of $\mathbb{R}^N$ that is just as good. For instance, in $\mathbb{R}^2$ we can form an orthogonal basis from $\mathbf{v}_1 = 2\mathbf{i} + \mathbf{j}$ and $\mathbf{v}_2 = -\mathbf{i} + 2\mathbf{j}$. Figure 8.1 shows that $\mathbf{w} = 3\mathbf{i} + 5\mathbf{j}$ is the sum of its orthogonal projections onto $\mathbf{v}_1$ and $\mathbf{v}_2$.

As we explained in Sect. 6.3, vectors acquire their components through basis expansion. In the standard basis $\{\mathbf{i}, \mathbf{j}\}$ the vector $\mathbf{w} = 3\mathbf{i} + 5\mathbf{j}$ has components $\begin{bmatrix} 3 & 5 \end{bmatrix}^T$. In the basis $\{\mathbf{v}_1, \mathbf{v}_2\}$, shown in Fig. 8.1,

$$\mathbf{w} = \frac{\mathbf{v}_1 \cdot \mathbf{w}}{\mathbf{v}_1 \cdot \mathbf{v}_1} \mathbf{v}_1 + \frac{\mathbf{v}_2 \cdot \mathbf{w}}{\mathbf{v}_2 \cdot \mathbf{v}_2} \mathbf{v}_2 = \frac{11}{5} \mathbf{v}_1 + \frac{7}{5} \mathbf{v}_2.$$

So, the components of $\mathbf{w}$ with respect to $\{\mathbf{v}_1, \mathbf{v}_2\}$ are $\begin{bmatrix} \frac{11}{5} & \frac{7}{5} \end{bmatrix}^T$. Ordinarily, computing vector components in a new basis requires solving a linear system of equations. However, if the basis is orthogonal, the components can be independently computed using the dot product.

8.2 Real and Complex DFT

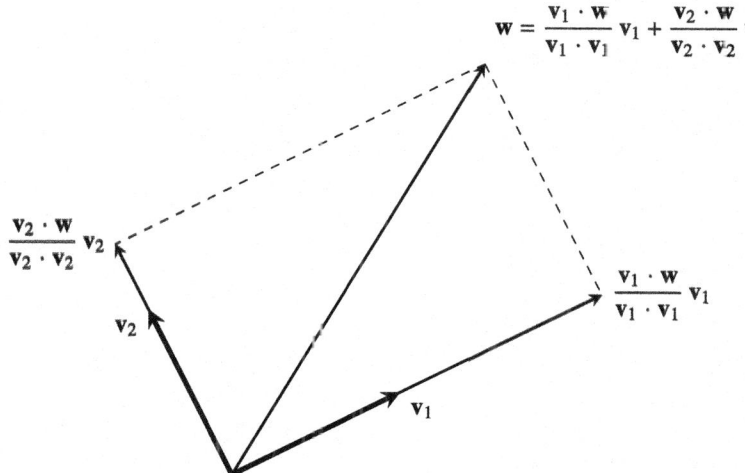

Fig. 8.1 Orthogonal expansion in $\mathbb{R}^2$: the vector $\mathbf{w} = 3\mathbf{i} + 5\mathbf{j}$ is the sum of its orthogonal projections onto $\mathbf{v}_1 = 2\mathbf{i} + \mathbf{j}$ and $\mathbf{v}_2 = -\mathbf{i} + 2\mathbf{j}$

8.2 Real and Complex DFT

Let $T > 0$, $\omega = 2\pi/T$, and let N be a positive integer. Cover the interval $[0, T]$ with the grid

$$t_k = T\frac{k-1}{N}, \quad k = 1, \ldots, N \tag{8.3}$$

and regard the grid points as components of the vector $\mathbf{t}$. It so happens that, when sampled on the grid (8.3), the constant function 1 and the functions $\cos(n\omega t)$ and $\sin(n\omega t)$ form an orthogonal basis of $\mathbb{R}^N$. Specifically, if N is odd then

$$\left\{ \mathbf{1}, \cos(\omega \mathbf{t}), \ldots, \cos\left(\frac{N-1}{2}\omega \mathbf{t}\right), \sin(\omega \mathbf{t}), \ldots, \sin\left(\frac{N-1}{2}\omega \mathbf{t}\right) \right\}$$

is an orthogonal basis (**1** is the vector of all 1's); whereas if N is even then

$$\left\{ \mathbf{1}, \cos(\omega \mathbf{t}), \ldots, \cos\left(\frac{N}{2}\omega \mathbf{t}\right), \sin(\omega \mathbf{t}), \ldots, \sin\left(\frac{N-2}{2}\omega \mathbf{t}\right) \right\}$$

is an orthogonal basis. For definiteness and simplicity, let us assume that N is odd. As follows from (8.2), if f is any function on $[0, T]$ then

$$f(\mathbf{t}) = a_0 \mathbf{1} + \sum_{n=1}^{\frac{1}{2}(N-1)} a_n \cos(n\omega \mathbf{t}) + b_n \sin(n\omega \mathbf{t}), \tag{8.4}$$

where
$$a_n = \frac{\cos(n\,\omega\,\mathbf{t}) \cdot f(\mathbf{t})}{\cos(n\,\omega\,\mathbf{t}) \cdot \cos(n\,\omega\,\mathbf{t})}, \quad b_n = \frac{\sin(n\,\omega\,\mathbf{t}) \cdot f(\mathbf{t})}{\sin(n\,\omega\,\mathbf{t}) \cdot \sin(n\,\omega\,\mathbf{t})}. \tag{8.5}$$

The coefficients (8.5), when arranged into a vector, form the DFT of the vector $f(\mathbf{t})$; we will also refer to them as the real DFT coefficients of the function f.

Equation (8.4) states that on the grid (8.3) the function f exactly matches the trigonometric polynomial
$$a_0 + \sum_{n=1}^{\frac{1}{2}(N-1)} a_n \cos(n\,\omega\,t) + b_n \sin(n\,\omega\,t). \tag{8.6}$$

If two functions agree on a grid, it is reasonable to expect them to be close in between grid points, especially if the grid is very fine. We can therefore regard the trigonometric polynomial (8.6) with coefficients given by (8.5) as a continuous approximation of f.

Let us investigate the DFT approximation of the function
$$f(t) = \begin{cases} 1, & \text{if } 0 \leq t \leq 1, \\ 0, & \text{otherwise.} \end{cases} \tag{8.7}$$

The following code computes DFT coefficients of (8.7) and its DFT approximation (8.6) for $T = 5$ and $N = 17$. This is a very inefficient way to compute DFT but it validates (8.5) in a transparent manner.

```
f  = @(t) double((t>=0) & (t<=1));
T  = 5; omega = 2*pi/T; N = 17;
t  = T*(0:N-1)/N; y = f(t);
tt = linspace(0,T,1000); yy = f(tt);
v  = ones(size(t)); a0 = dot(v,y)/dot(v,v);
zz = a0*ones(size(tt));
a  = zeros(1,.5*(N-1)); b = a;
for n=1:.5*(N-1)
    v = cos(n*omega*t); a(n) = dot(v,y)/dot(v,v);
    zz = zz + a(n)*cos(n*omega*tt);
    v = sin(n*omega*t); b(n) = dot(v,y)/dot(v,v);
    zz = zz + b(n)*sin(omega*n*tt);
end
figure;
subplot(2,2,1); stem(0:.5*(N-1),[a0 a],'k');
xlim([-1 .5*(N-1) + 1]); ylim([-.1 .4]);
text(4,.3,'a-coefficients');
subplot(2,2,2); stem(1:.5*(N-1),b,'k');
xlim([0 .5*(N-1) + 1]); ylim([-.1 .4]);
text(4,.3,'b-coefficients')
subplot(2,2,3:4); plot(tt,yy,'k-',t,y,'ko',tt,zz,'k-');
xlim([0 T]); xlabel('t');
text(T-3,1,'DFT approximation');
```

8.2 Real and Complex DFT

The results are displayed in Fig. 8.2. The bottom plot shows that the trigonometric polynomial (8.6) not only exactly matches the rectangle function (8.7) on the grid, but also approximates it on the entire interval.

Figure 8.3 shows what happens if the grid in Fig. 8.2 is refined: the DFT approximation of (8.7) gets better as $N \to \infty$. It also shows that the approximation is not uniform. Since sines and cosines are smooth, so is the trigonometric polynomial which, for that reason, struggles to reproduce jump discontinuities. This manifests as characteristic overshoots known as the *Gibbs phenomenon*.

If the function is continuous or, better yet, differentiable, then there is no Gibbs phenomenon and the convergence of the DFT approximation is much faster. Figure 8.4 illustrates that for the functions

$$f(t) = T - |2t - T| \tag{8.8}$$

and

$$f(t) = T\,e^{-(2t-T)^2}, \tag{8.9}$$

with $T = 5$ and $N = 17$. The function (8.8) (left panel of Fig. 8.4) is continuous throughout $[0, T]$ but not differentiable; notice how the trigonometric polynomial rounds the sharp corner of (8.8)—this rounding can be diminished by increasing N. The gaussian (8.9) is not differentiable at the endpoints of $[0, T]$ but it decreases so rapidly away from the center of the interval that it behaves like a smooth function. As a result, the right panel of Fig. 8.4 appears to have only one plot because (8.9) is indistinguishable from the trigonometric polynomial at this scale.

Together Figs. 8.3 and 8.4 suggest that, as $N \to \infty$, the trigonometric polynomial (8.4) converges to the function f, which is indeed the case.

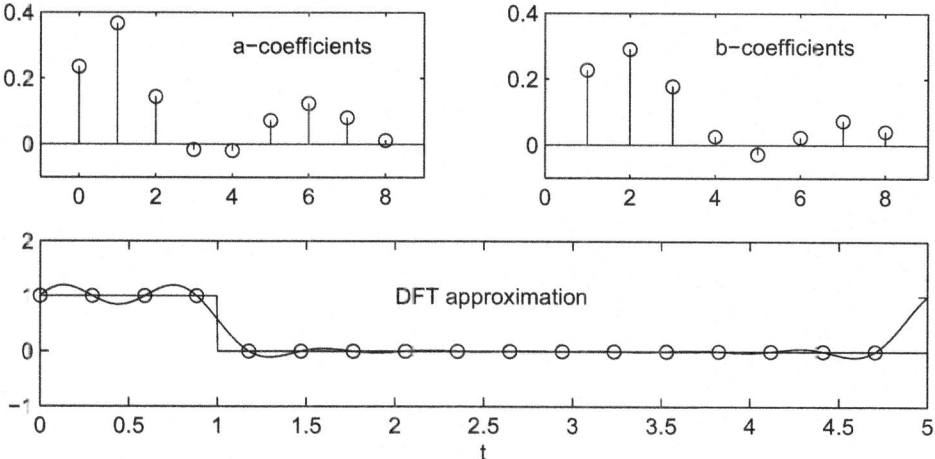

Fig. 8.2 DFT approximation of (8.7) with $T = 5$ and $N = 17$

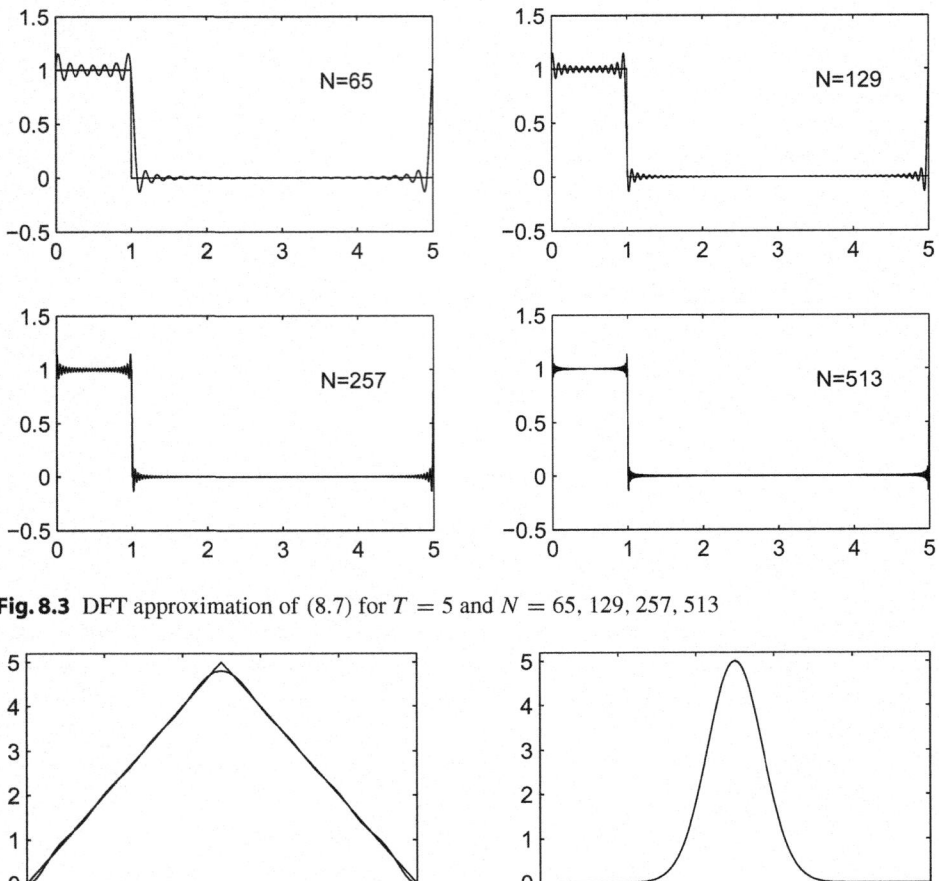

Fig. 8.3 DFT approximation of (8.7) for $T = 5$ and $N = 65, 129, 257, 513$

Fig. 8.4 DFT approximations of (8.8) (left panel) and (8.9) (right panel) with $T = 5$ and $N = 17$

As we have seen in Chap. 5, and, more recently, in Sect. 7.4, computations involving real harmonics tend to simplify if the latter are replaced with complex exponentials. This is true of the DFT which is commonly implemented in complex form.

Before formulating complex DFT, we must define orthogonality in $\mathbb{C}^N$. The dot product is not suitable for that—we need complex *inner product*

$$\langle \mathbf{w}, \mathbf{v} \rangle = \sum_{k=1}^{N} \overline{w_k}\, v_k \tag{8.10}$$

8.2 Real and Complex DFT

which looks like the dot product but involves conjugation for reasons that will become clear in Sect. 9.2. Just as two vectors in $\mathbb{R}^n$ are orthogonal if their dot product is zero, two vectors $\mathbf{w}$ and $\mathbf{v}$ in $\mathbb{C}^n$ are orthogonal if their complex inner product is zero: $\langle \mathbf{w}, \mathbf{v} \rangle = 0$.

The following computation shows that complex exponentials $e^{in\omega t}$, when sampled on the grid (8.3), form an orthogonal basis of $\mathbb{C}^N$ with respect to the complex inner product (8.10):

$$\langle e^{in\omega t}, e^{im\omega t} \rangle = \sum_{k=1}^{N} \overline{e^{in\omega t_k}} \, e^{im\omega t_k} = \sum_{k=1}^{N} e^{i(m-n)\omega t_k}$$

$$= \sum_{k=1}^{N} \left(e^{\frac{2\pi i (m-n)}{N}} \right)^{k-1}.$$

If $n \neq m$ the geometric sum vanishes

$$\sum_{k=1}^{N} \left(e^{\frac{2\pi i (m-n)}{N}} \right)^{k-1} = \frac{1 - e^{2\pi i (m-n)}}{1 - e^{\frac{2\pi i (m-n)}{N}}} = 0,$$

since $e^{2\pi i (m-n)} = 1$, according to Euler's formula. Thus, if $n \neq m$, $e^{in\omega t}$ is orthogonal to $e^{im\omega t}$; if $m = n$ then

$$\sum_{k=1}^{N} \left(e^{\frac{2\pi i (n-n)}{N}} \right)^{k-1} = \sum_{k=1}^{N} 1 = N.$$

This shows that the square of the norm of $e^{in\omega t}$ equals N regardless of n.

Having shown orthogonality of $e^{in\omega t}$ we now define complex DFT coefficients of $f(\mathbf{t})$ by the relation

$$c_n = \frac{\langle e^{in\omega t}, f(\mathbf{t}) \rangle}{\langle e^{in\omega t}, e^{in\omega t} \rangle} = \frac{1}{N} \langle e^{in\omega t}, f(\mathbf{t}) \rangle. \tag{8.11}$$

The range of n depends on the parity of N: if N is odd then $n = -\frac{N-1}{2}, \ldots, \frac{N-1}{2}$; if N is even then $n = -\frac{N-2}{2}, \ldots, \frac{N}{2}$.

Assuming again that N is odd, for simplicity, we can write the complex analog of (8.4) as

$$f(\mathbf{t}) = \sum_{n=-\frac{N-1}{2}}^{\frac{N-1}{2}} c_n e^{in\omega t}, \quad c_n = \frac{1}{N} \langle e^{in\omega t}, f(\mathbf{t}) \rangle \tag{8.12}$$

and reason, as before, that exact equality on the grid must force some degree of agreement throughout the entire interval:

$$f(t) \approx \sum_{n=-\frac{N-1}{2}}^{\frac{N-1}{2}} c_n e^{in\omega t}. \tag{8.13}$$

That is, indeed, the case. In fact, approximate equality in (8.13) becomes exact as $N \to \infty$.

As we will now show through examples, many signals can be approximated with sums of the form (8.13) where summation extends over a relatively small range $|n| \leq b$ encompassing only those complex harmonics whose coefficients have significant magnitudes. Approximating signals with sums of dominant complex harmonics is the primary use of DFT in data analysis.

8.3 Filtering Out Noise

This example uses artificial data to demonstrate MATLAB's fft which implements the fast Fourier transform algorithm. The routine computes complex DFT coefficients (8.11), but it also scales them by the number of data points and returns them in a specific order. For instance, if **x** is a row vector of length 8 and c_n are the complex DFT coefficients (8.11) with $n = -3, \ldots, 4$ the output of fft(x) is the row vector

$$8 \begin{bmatrix} c_0 \ c_1 \ c_2 \ c_3 \ c_4 \ c_{-3} \ c_{-2} \ c_{-1} \end{bmatrix}.$$

For real-valued data the complex Fourier coefficients have symmetry: $c_n = \overline{c_{-n}}$. Since our data will be real-valued, we will work with the first half of the coefficients returned by fft which we will divide by N.

The data shown in Fig. 8.5 is a sum of three harmonics contaminated by white noise:
```
T = 50; N = 16384; t = T*(0:N-1)/N;
x = .5*cos(2*pi*t) + 1.5*cos(4*pi*t + pi/3) + ...
    .75*cos(6*pi*t) + randn(size(t));
figure; plot(t,x,'k-'); xlim([0 10]); ylim([-7 7]);
```

We intentionally shortened the limits of the x-axis of Fig. 8.5 so that periodicity is more apparent. Periodicity of the data almost always calls for DFT processing.

To clean up the data, we first need to analyze its frequency content. To this end, we plot twice the magnitude of DFT coefficients c_n for $n \geq 0$ (the factor of 2 is not essential: it

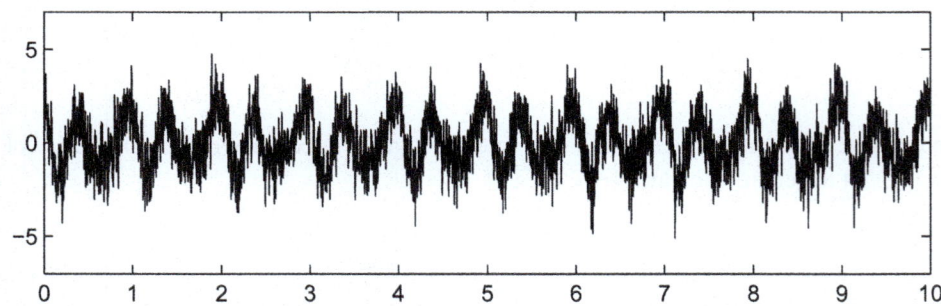

Fig. 8.5 Sum of three harmonics contaminated by white noise

8.3 Filtering Out Noise

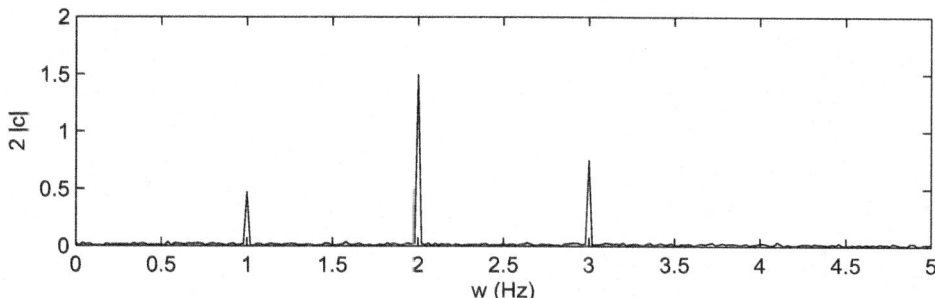

Fig. 8.6 Frequency content of the data in Fig. 8.5

accounts for the fact that we ignore c_n for negative n). The following code actually plots $2|c_n|$ for all n in Fig. 8.6, but we shorten the x-axis to focus on the part of the spectrum that is most relevant.

```
w = (0:N-1)/T; y = fft(x)/N; z = 2*abs(y);
figure; plot(w,z,'k-');
xlim([0 5]); ylim([0 2]);
xlabel('w (Hz)'); ylabel('2 |c|');
```

The three sharp peaks in Fig. 8.6 tell us that the data is a sum of three harmonics (there are no other peaks in the plot); the centers of the peaks give the frequencies while the heights give the amplitudes.

Now, to remove the noise, we set all but six of the largest DFT coefficients (two per peak) to zero and apply inverse DFT:

```
[z,ind] = sort(z,'descend');
y(ind(7:end)) = 0;
z = N*ifft(y);
figure; plot(t,z,'k-'); xlim([0 10]); ylim([-3 3]);
```

Figure 8.7 shows the data with the noise removed. The amplitudes and phases of the six dominant harmonics can be extracted as follows:

```
ind = ind([1 3 5]);
w0  = w(ind);             % frequency
c0  = y(ind);
a0  = 2*abs(c0);          % amplitude
phi0 = angle(c0);         % phase
```

Figure 8.8 compares the parameters of dominant harmonics extracted by DFT with those used to generate data. The frequencies are determined exactly while the amplitudes and phases are in close agreement.

Of course, for data collected in the field it is unlikely that we would know the exact parameters of the main harmonic constituents. This makes statistical analysis of the residual particularly important.

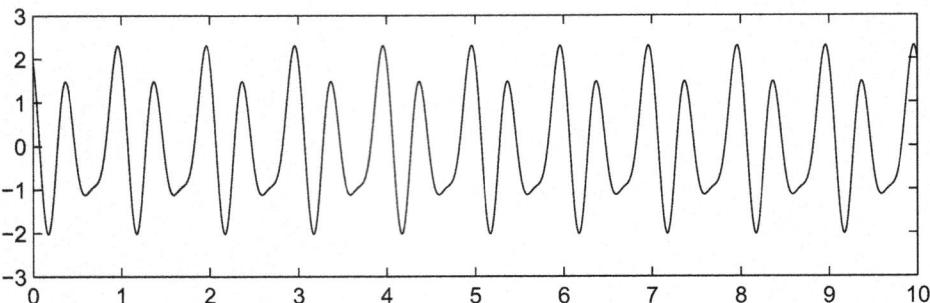

Fig. 8.7 Data in Fig. 8.5 with the noise removed

Frequency		Amplitude		Phase	
Exact	DFT	Exact	DFT	Exact	DFT
1	1	0.5	0.4687	0	0.0132
2	2	1.5	1.4950	1.0472	1.0424
3	3	.75	0.7480	0	0.0328

Fig. 8.8 Frequency content of filtered data and data not contaminated by noise

The following code produces Fig. 8.9.

```
r = x - z; m = mean(r); v = var(r);
s = linspace(-5,5,50); ds = s(2)-s(1);
h = hist(r,s)/ds/N;
figure;
subplot(1,2,1);
plot(t,r,'k-'); xlim([0 10]);
subplot(1,2,2);
bar(s,h,'barwidth',1,'facecolor','w');
hold on;
plot(s,exp(-.5*s.^2)/sqrt(2*pi),'k-');
legend('residual','N(0,1)'); xlim([-5 5]);
```

According to Fig. 8.9, the residual is white noise—the computed mean 0.0096 and variance 1.0194 are very close to the parameters of the standard normal distribution $N(0, 1)$ that was used to contaminate the data. Since the residual is white noise, all of the important harmonic constituents have been successfully extracted.

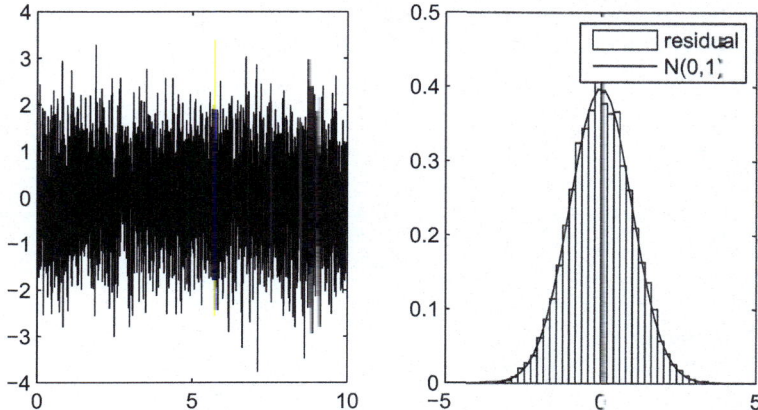

Fig. 8.9 Residual and its histogram compared to $N(0, 1)$

8.4 Comparison of Tides

Figure 8.10 shows January 2022 tide level (with the mean subtracted) recorded by NOAA stations 9411340 (Santa Barbara, CA) and 8413320 (Bar Harbor, ME).

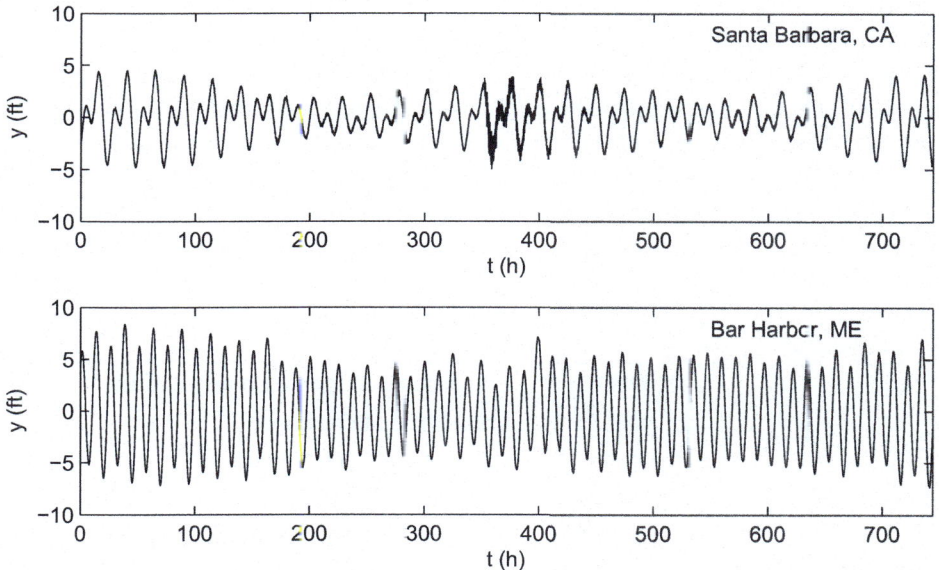

Fig. 8.10 Tide level for Santa Barbara, CA (top panel) and Bar Harbor, ME (bottom panel) in January 2022 with the mean subtracted. (NOAA Tides and Currents)

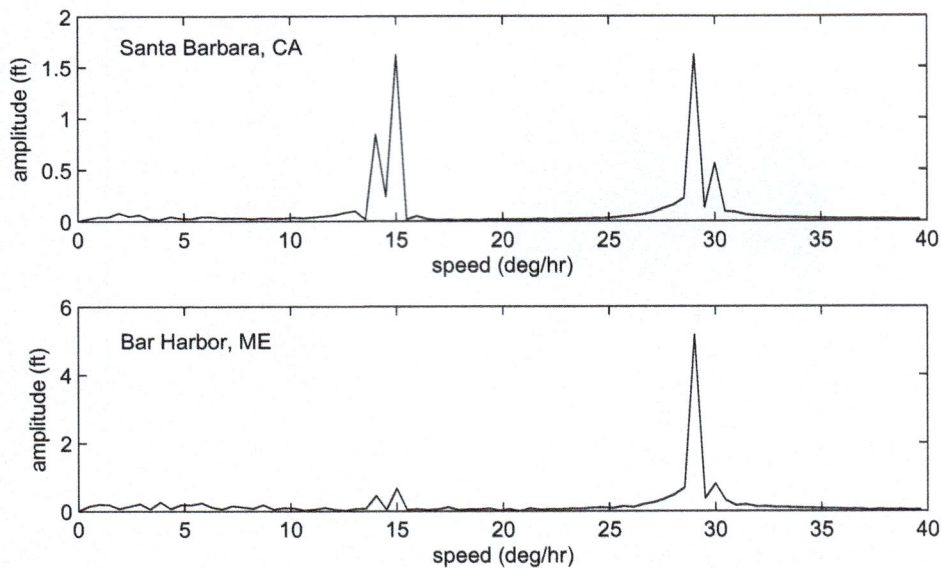

Fig. 8.11 Frequency content of the data in Fig. 8.10

The two data sets look very different; after all, these are tides in two different oceans. Yet Fig. 8.11 reveals that five out of the first six dominant harmonics in both data sets have the same frequencies, or speeds, as NOAA calls them.

The perfect alignment of the peaks in Fig. 8.11 is not a coincidence. The two main driving forces causing tides are the gravitational forces exerted on the oceans by the moon and the sun. As the moon rotates around the earth and the earth rotates around the sun, while spinning around its own axis, these two forces interact in a complicated manner, but in such a way that all tides have dominant harmonic components of certain speeds. Figure 8.12 lists six largest harmonic constituents for tides in Santa Barbara, CA with speeds determined by NOAA.

Constituent	Speed ($° h^{-1}$)	Description
Q_1	13.398661	Larger lunar elliptic diurnal constituent
O_1	13.943035	Lunar diurnal constituent
K_1	15.041069	Lunar diurnal constituent
M_2	28.984104	Principal lunar semidiurnal constituent
S_2	30.0	Principal solar semidiurnal constituent
N_2	28.43973	Larger lunar elliptic semidiurnal constituent

Fig. 8.12 Six dominant harmonic constituents for tides in Santa Barbara (NOAA)

8.5 Forecasting Solar Activity

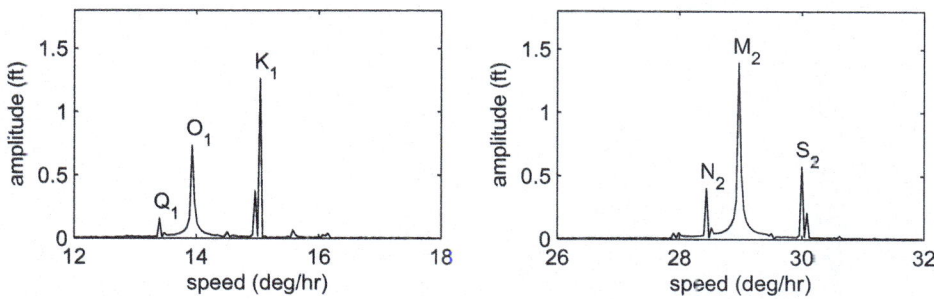

Fig. 8.13 Frequency content of tide data for Santa Barbara, CA for calendar year 2022 (Tide data provided by NOAA Tides and Currents)

The corresponding spectral peaks are identified in Fig. 8.13 which was computed using NOAA data for the entire calendar year of 2022.

8.5 Forecasting Solar Activity

Figure 8.14 shows daily sunspot data and its frequency content.

The largest component has frequency $0.0909\,\text{year}^{-1}$ corresponding to the period of 11 years; the next three significant components have periods of 33, 8.25 and 5.5 years. In order

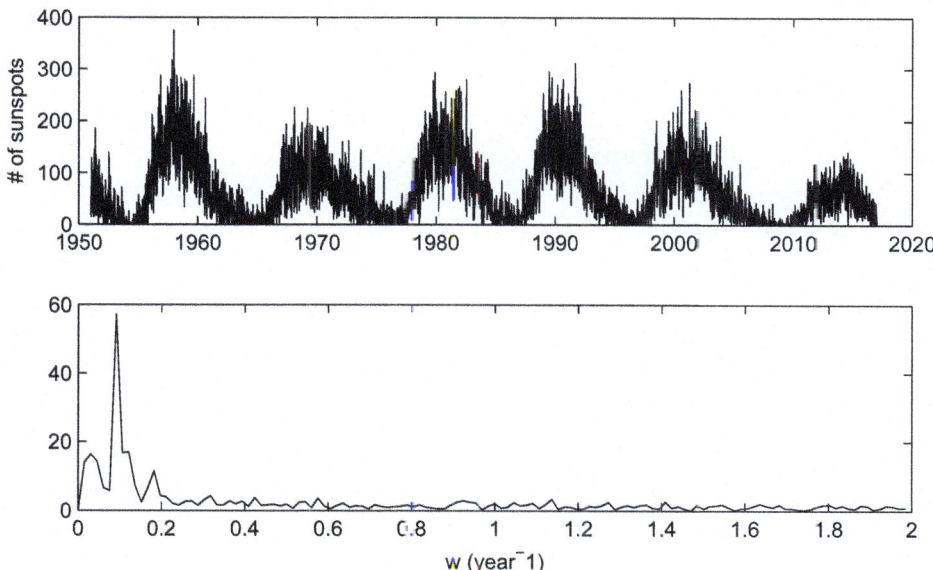

Fig. 8.14 NOAA sunspot data (top) and its frequency content (bottom) computed after the mean is subtracted

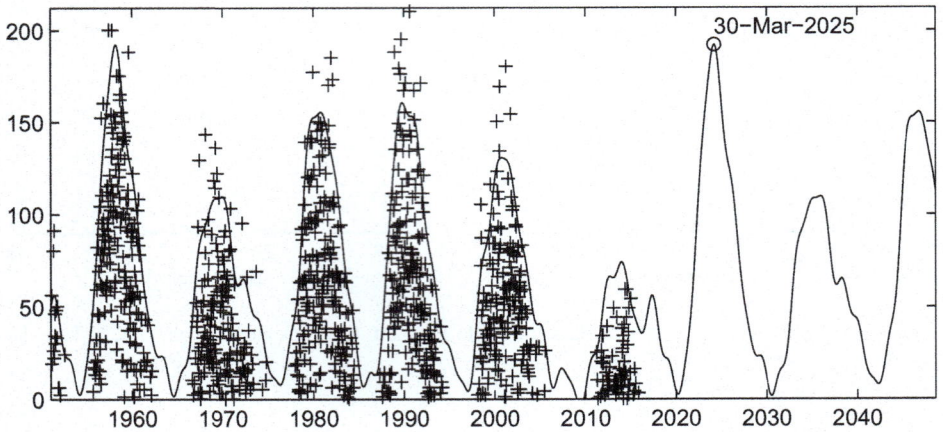

Fig. 8.15 Forecasting solar activity using 32 dominant harmonic components of the sunspot data in Fig. 8.14

to forecast solar activity, we decided to use 32 dominant harmonic components. Figure 8.15 shows every tenth observation from the sunspot data set (crosses) and the DFT approximation (solid line). According to the figure, the next peak of solar activity (with 192 sunspots) should be expected on March 30, 2025.

Of course, the predicted date is very approximate. If we use more harmonics or fewer harmonics, the prediction can shift by several months. Still, it is fair to say, based on the data, that we are approaching peak solar activity and that it will happen either in late 2024 or in 2025.

8.6 DFT Solution of an ODE

As a final example, we will use DFT to approximate the solution of the following IVP

$$\frac{dx}{dt} + k\, x = f(t), \quad x(0) = x_0 \tag{8.14}$$

on the interval $[0, T]$. The idea is to approximate the right-hand side with a sum of complex harmonics

$$f \approx \sum_n c_n\, e^{i n \omega t}, \quad \omega = \frac{2\pi}{T} \tag{8.15}$$

and use the principle of superposition. The particular solution corresponding to (8.15) is

$$x_p = \sum_n \frac{c_n}{i n \omega + k}\, e^{i n \omega t}. \tag{8.16}$$

The solution of IVP (8.14) therefore is approximately,

$$x = (x_0 - x_p(0))\, e^{-kt} + x_p(t).$$

Notice that the nth DFT coefficient of the particular solution (8.16) is the nth DFT coefficient of the right-hand side f divided by $(in\omega + k)$. This means that we can generate x_p by computing DFT coefficients of f, dividing them by appropriate factors, and applying the inverse DFT, as we do in the following code.

```
f = @(t) double(t<=1); k = .75;
odefun = @(t,x) -k*x + f(t); x0 = -.25;
T = 5; N = 1024; t = T*(0:N-1)/N;
opts = odeset('RelTol',1e-6);
[t,x] = ode45(odefun,t,x0,opts);
c = fft(f(t));
w = [0:.5*N -.5*(N-2):-1]'/T;
xp = ifft(c./(k+1i*2*pi*w));
xc = (x0-xp(1))*exp(-k*t);
figure
plot(t(1:20:end),x(1:20:end),'ko');
hold on;
plot(t,real(xc + xp),'k-');
xlabel('t'); ylabel('x');
legend('ode45','DFT')
```

The output is shown in Fig. 8.16. In the code we set the right-hand side f of the IVP (8.14) to (8.7), but that choice was arbitrary and can be changed. We also decreased the relative error tolerance RelTol to force ode45 to compute the solution more accurately; with the default value of RelTol (which is 1×10^{-3}) ode45 produces large errors because of the jump discontinuity in f.

8.7 Comments and Bibliography

One of the first DFT's was computed by Gauss to determine the orbit of an asteroid named Ceres. Actually, in the process of computing DFT, Gauss discovered its fast version FFT but, as was his wont, did not give it any publicity. Gauss's FFT algorithm lay dormant for more than two centuries until it was rediscovered in 1965 by Cooley and Tukey. That was when the field of digital signal processing was born.

DFT, being a linear operator on $\mathbb{R}^N$, can be implemented as matrix multiplication. However, straightforward multiplication of a vector in $\mathbb{R}^N$ by an N-by-N matrix has complexity $O(N^2)$. Meanwhile, Gauss's algorithm, rediscovered by Cooley and Tukey, has complexity of $O(N \log N)$ for N that is a power of 2. If N is not a power of 2, there are various tricks of maintaining $O(N \log N)$ complexity. However, if given a choice, it is always best

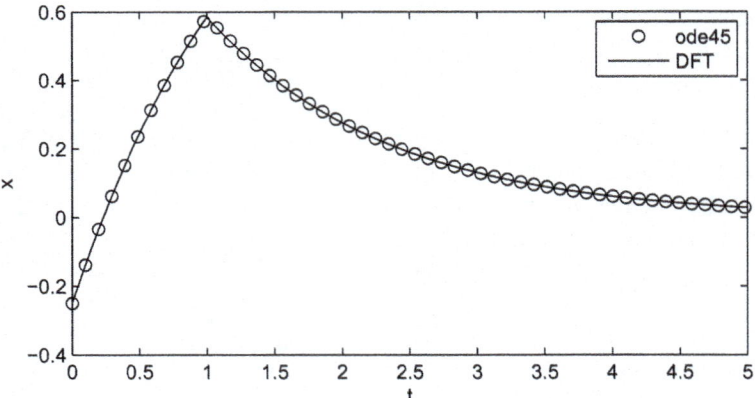

Fig. 8.16 DFT solution of IVP (8.14) with $k = 0.75$, $x_0 = -0.25$ and the right-hand side set to (8.7). The solution was computed on [0,5] using a grid with 1024 points; to avoid clutter, every 20th value computed by `ode45` is shown

to perform DFT processing on inputs that have power-of-2 length. For more on DFT/FFT see [1].

Harmonic tide analysis was developed by Sir William Thomson (Lord Kelvin) around 1867. Sir Thomson also designed a tide-predicting machine, a mechanical computer that was built in 1872. A dated, but very readable account of tide theory and its history can be found in [4].

There is some disagreement among the experts about Solar Cycle 25 whose peak we predicted to occur on March 30, 2025. NOAA thinks that the peak will be in June of that year, but with a margin of error of ±5 months. The authors of [3] think that Solar Cycle 25 will peak in mid-2024 (see also [2]).

8.8 Exercises

1 Let $\{\mathbf{v}_k\}_{k=1}^n$ be a collection of mutually orthogonal vectors in $\mathbb{R}^N$. In this problem we assume that $n < N$, so $\{\mathbf{v}_k\}_{k=1}^n$ is not a basis of $\mathbb{R}^N$. While we cannot necessarily represent every vector $\mathbf{w} \in \mathbb{R}^N$ with a linear combination of $\mathbf{v}_k$'s, we can consider the approximation

$$\mathbf{w} \approx \sum_{k=1}^n c_k \, \mathbf{v}_k. \tag{8.17}$$

Find the coefficients c_k in (8.17) by minimizing the 2-norm of the residual $\|\mathbf{w} - \sum_{k=1}^n c_k \, \mathbf{v}_k\|$. What does this tell you about orthogonal approximations?

8.8 Exercises

2. Download NOAA tide data for a station of your choice and identify as many harmonic constituents as possible. You will need at least a month worth of observations but the more data the better.

3. Use DFT to solve the following IVP describing a 1DOF mass-spring system:

$$m\ddot{x} + r\dot{x} + kx = f, \quad x(0) = \dot{x}(0) = 0.$$

Experiment with the parameters m, r, and k, the forcing term f, the interval $[0, T]$, and the number of grid points N; validate your answer using either an exact formula or the output of ode45. Bear in mind that, depending on the size of the interval and the behavior of the forcing term, you may need to tighten the relative error tolerance to ensure that ode45 computes an accurate solution.

4. Consider a solar system with a single planet orbiting a sun of mass M. Let $\mathbf{r}(t) = \begin{bmatrix} x(t) & y(t) \end{bmatrix}^T$ be the position of the planet relative to the sun, which is placed at the origin. According to Newton's second law and his law of universal gravitation

$$\ddot{\mathbf{r}} = -\frac{GM}{\|\mathbf{r}\|^3} \mathbf{r}, \qquad (8.18)$$

where G is the universal gravitational constant. The following function solves (8.18) and plots the coordinates of the planet:

```
function kepler
GM = 1;
    function du = odefun(t,u)
        r = sqrt(u(1)^2+u(3)^2);
        s = -GM/r^3;
        du = [u(2); s*u(1); u(4); s*u(3)];
    end
u0 = [1; -.5; 1; 1];
options = odeset('RelTol',1e-06,'AbsTol',1e-6);
[t,u] = ode45(@odefun,0:1023,u0,options);
figure;
subplot(2,1,1); plot(t,u(:,1),'k-');
xlim([0 1023]); xlabel('t'); ylabel('x');
subplot(2,1,2); plot(t,u(:,3),'k-');
xlim([0 1023]); xlabel('t'); ylabel('y');
end
```

In the code we set $GM = 1$ and chose initial conditions corresponding to a typical elliptical orbit. Generate the data shown in Fig. 8.17 and use it to find the following:

a. The length of the planetary year.
b. The lengths of the semi-minor and semi-major axes of the planetary orbit.

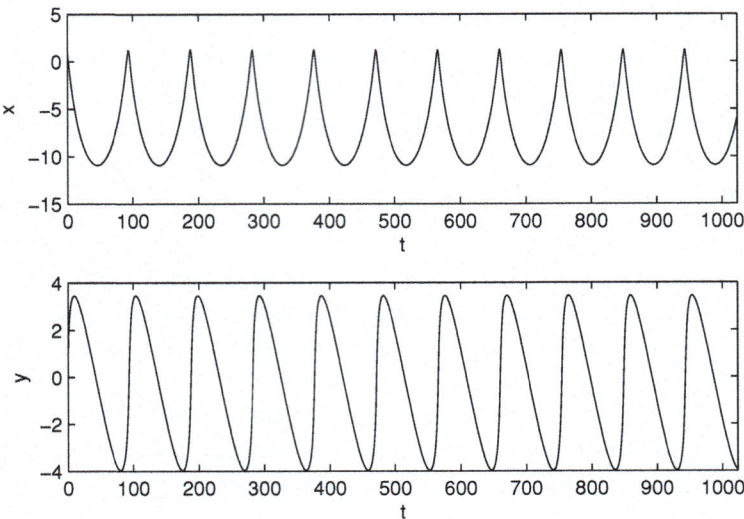

Fig. 8.17 Solution of (8.18)

A more interesting version of this problem would involve actual astronomical observations. However, it would then require much deeper knowledge of astronomy and celestial mechanics.

References

1. W.L. Briggs, van E. Henson, *The DFT: an Owner's Manual for the Discrete Fourier Transform* (Society for Industrial and Applied Mathematics, Philadelphia, 1995)
2. F. Clette, E.W. Cliver, L. Lefèvre, L. Svalgaard, J.M. Vaquero, Revision of the sunspot number(s). Space Weather **13**(9), 529–530 (2015)
3. S. Mcintosh, S. Chapman, R. Leamon, R. Egeland, N. Watkins, Overlapping magnetic activity cycles and the sunspot number: forecasting sunspot cycle 25 amplitude. Solar Phys. **295**, 12 (2020)
4. P. Schureman, Manual of harmonic analysis and prediction of tides, in *US Coast and Geodetic Survey, Special Publication 98* (United States Government Printing Office, Washington DC, 1958). Revised 1940 edition reprinted 1958 with corrections, reprinted 2001

Fourier Series

In Sect. 8.2 we showed that a function f can be approximated on the interval $[0, T]$ with a linear combination of complex exponentials whose (circular) frequencies are integer multiples of $\omega = 2\pi/T$ and whose coefficients are given by DFT. Let the interval $[0, T]$ be covered by the grid (8.3), as in Sect. 8.2, and let N be odd so that the complex DFT approximation is given by (8.12). We will now show that, for a reasonable function f, the limit of (8.12) as $N \to \infty$ exists and equals the *Fourier value* of the function:

$$\lim_{N \to \infty} \sum_{n=-\frac{N-1}{2}}^{\frac{N-1}{2}} c_n e^{in\omega t} = \frac{f(t+) + f(t-)}{2}, \quad f(t\pm) = \lim_{\tau \to t^{\pm}} f(\tau). \tag{9.1}$$

By "reasonable" we mean a function that has finitely many jump discontinuities and no other singularities. Notice that if the function is continuous then its value is the same as its Fourier value.

As $N \to \infty$ the sum in (9.1) becomes the Fourier series of f with summation extending from $-\infty$ to ∞. Meanwhile, the DFT coefficients c_n become Fourier series coefficients $\widehat{f_n}$:

$$\widehat{f_n} = \lim_{N \to \infty} c_n. \tag{9.2}$$

In Sect. 9.1 we will identify the limits (9.2) as certain integrals and numerically confirm that a function whose values coincide with its Fourier values has Fourier series representation

$$f(t) = \sum_{n=-\infty}^{\infty} \widehat{f_n} e^{in\omega t}. \tag{9.3}$$

A highly technical proof of (9.3) is omitted, but an interested reader will find it in references cited in the Comments and Bibliography section at the end of the chapter.

© The Author(s), under exclusive license to Springer Nature Switzerland AG 2025
A. Beltukov, *Differential Equations and Data Analysis*, Synthesis Lectures on Mathematics & Statistics, https://doi.org/10.1007/978-3-031-62257-1_9

In Chap. 8 (Sect. 8.6) we used DFT to approximate the solution of a first order scalar linear ODE with constant coefficients. In Sect. 9.3 we will derive the exact solution of that ODE as a Fourier series. We will then show that constructing Fourier series solutions of higher order ODE and matrix-vector ODE is just as simple.

9.1 Representation of Functions

To unravel the limits (9.2), we first invoke the definition of the complex DFT coefficients (8.11) in terms of the complex inner product (8.10):

$$\lim_{N\to\infty} c_n = \lim_{N\to\infty} \frac{1}{N} \langle e^{in\omega \mathbf{t}}, f(\mathbf{t}) \rangle = \lim_{N\to\infty} \frac{1}{N} \sum_{k=1}^{N} e^{-in\omega t_k} f(t_k). \tag{9.4}$$

The sum in (9.4) closely resembles a Riemann sum. To recognize it as such, it is helpful to rewrite the limit (9.4) as

$$\frac{1}{T} \lim_{N\to\infty} \sum_{k=1}^{N} e^{-in\omega t_k} f(t_k) \frac{T}{N}.$$

The quantity T/N is the spacing of the grid $\mathbf{t}$. Hence the sum inside the limit is the left endpoint Riemann sum for the integral of $e^{-in\omega t} f(t)$ over $[0, T]$. We conclude that the Fourier series coefficients are given by

$$\widehat{f_n} = \frac{1}{T} \int_0^T e^{-in\omega t} f(t)\, dt. \tag{9.5}$$

The integral in (9.5) is, essentially, a continuous limit of the discrete complex inner product (8.10). It therefore makes sense to regard it as a continuous inner product of f and the complex exponential: $\widehat{f_n} = \langle e^{in\omega t}, f(t) \rangle$. As will be explained in Sect. 9.2, the inner product of complex-valued functions on $[0, T]$ is generally defined by

$$\langle g, f \rangle = \int_0^T \overline{g(t)}\, f(t)\, dt. \tag{9.6}$$

Setting that aside for the time being, let us test (9.5) numerically by approximating a few functions with partial sums of their Fourier series (9.3). Specifically, let us examine the dependence of the error of Fourier series approximation

$$e_N(f) = \max_{0\le t\le T} |f(t) - \sum_{n=-N}^{N} \widehat{f_n} e^{in\omega t}| \tag{9.7}$$

on the *bandwidth N*.

9.1 Representation of Functions

In what follows, we will determine the rate at which $e_N(f) \to 0$ as $N \to \infty$ for the following three test functions that were chosen to have varying degrees of smoothness (differentiability):

$$f_1(t) = \begin{cases} 1, & 0 < t < 1, \\ 0, & 1 < t < T, \\ 1/2, & t = 0, 1, T. \end{cases}$$

$$f_2(t) = T - |2t - T|, \qquad (9.8)$$

$$f_3(t) = \begin{cases} t^2, & 0 \leq t < T/3, \\ T^2/6 - 2(t - T/2)^2, & T/3 \leq t \leq 2T/3, \\ (t - T)^2, & 2T/3 < t \leq T. \end{cases}$$

The piecewise continuous function f_1 is from Sect. 8.2, however, we adjusted Eq. (8.7) so that f_1 returns its Fourier value $1/2$ at the jump discontinuities $t = 0, 1, T$. The continuous, but not differentiable function f_2 is also from Sect. 8.2 where it was defined by (8.8). The last function f_3 has one continuous derivative and is the smoothest of the three.

The Fourier series coefficients of the functions (9.8) are given in (9.9) (as functions of the integer index n):

$$\widehat{f_1}(n) = \begin{cases} (1 - e^{-i\omega n})/(i\omega n T), & n \neq 0, \\ 1/T, & n = 0. \end{cases}$$

$$\widehat{f_2}(n) = \begin{cases} -2\left(1 - e^{-i\omega n T/2}\right)^2/(\omega^2 n^2 T), & n \neq 0, \\ T/2, & n = 0. \end{cases} \qquad (9.9)$$

$$\widehat{f_3}(n) = \begin{cases} 2i\left(1 - e^{-i\omega n T/3}\right)^3/(\omega^3 n^3 T), & n \neq 0, \\ 2T/2/27, & n = 0. \end{cases}$$

We leave the computation of (9.9) as a Calculus exercise.

Figure 9.1 shows the log-log plots of (9.7) for the functions (9.8).

To produce Fig. 9.1, we implemented (9.8) and (9.9) as subfunctions at the start of the main function file.

```
T = 5;   w = 2*pi/T;

function y1 = f1(t)
y1 = double(t < 1);
ind = (t==0) | (t==1)| (t==T);
y1(ind,1) = .5;
end
```

```
function c1 = fhat1(n)
c1 = (1 - exp(-1i*w*n))./(1i*w*n*T);
c1(n==0) = 1/T;
end

function y2 = f2(t)
y2 = T - abs(2*t-T);
end

function c2 = fhat2(n)
c2 = -2*(1-exp(-.5*1i*T*w*n)).^2./(w^2*n.^2*T);
c2(n==0) = .5*T;
end

function y3 = f3(t)
y3 = t.^2;
ind = (t > T/3) & (t <= 2*T/3);
y3(ind) = T^2/6 - 2*(t(ind) - T/2).^2;
ind = (t > 2*T/3);
y3(ind) = (t(ind) - T).^2;
end

function c3 = fhat3(n)
c3 = 2*1i*(1-exp(-1i*T*w*n/3)).^3./(w^3*n.^3*T);
c3(n==0) = 2*T^2/27;
end
```

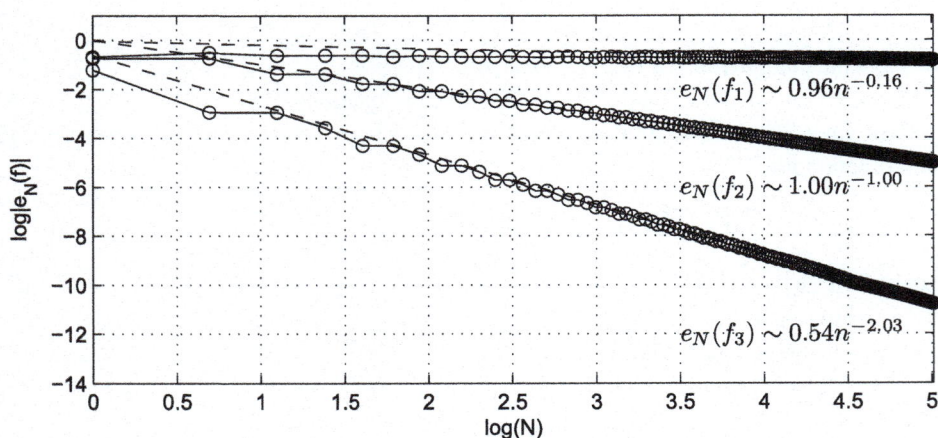

Fig. 9.1 Log-log plot of Fourier approximation error (9.7) for the functions (9.8)

We then sampled (9.8) on a fine grid, computed Fourier series coefficients (9.9) for $n = -150, \ldots, 150$, and pre-computed the corresponding complex exponentials as columns of a matrix with the following code.

```
t = linspace(0,T,1000)';   N = 150;  n = (-N:N)';
x = [f1(t)  f2(t)  f3(t)];
c = [fhat1(n) fhat2(n) fhat3(n)];
A = exp(1i*w*t*n');
```

Next, the errors (9.7) were computed in the loop, using matrix multiplication.

```
e = zeros(N,3);
for b=1:N
    ind = (abs(n) <= b);
    for k=1:3
        y = A(:,ind)*c(ind,k);
        e(b,k) = norm(abs(x(:,k) - y),inf);
    end
end
u = log(1:N)';  v = log(e);
figure; plot(u,v,'ko-'); hold on;
xlim([min(u) max(u)]); ylim([-14 1]);
xlabel('log(N)'); ylabel('log|e_N(f)|'); grid on;
```

Finally, we performed linear regression on the tail of the error data using MATLAB's polyfit command.

```
xpos = 3.5; ypos = [-2 -6 -12];
s = '$e_{N}(f_{%g})\\sim %1.2f n^{%1.2f}$';
for k=1:3
    p = polyfit(u(N-10:N),v(N-10:N,k),1);
    plot(u,polyval(p,u),'k--')
    text(xpos,ypos(k),sprintf(s,k,exp(p(2)),p(1)),...
        'interpreter','latex','fontsize',12)
end
```

It is clear from Fig. 9.1 that for the continuous function f_2 the error of the Fourier series approximation approaches zero at a linear rate, whereas for the once-differentiable function f_3 it decreases quadratically. As shown in an exercise at the end of the chapter, for a twice-differentiable function the error (9.7) decreases cubically. It is an easy guess that for a k-times differentiable function the Fourier error (9.7) decays as $N^{-(k+1)}$, which is indeed the case. If a function happens to have derivatives of all orders then the error decreases faster than any polynomial.

Figure 9.1 also shows that if a function has jump discontinuities, as does f_1, its Fourier series converges rather reluctantly. This is due to Gibbs phenomenon which we encountered in Sect. 8.2. Actually, if instead of estimating the error on a grid we measured the precise error on the entire interval, we would have found that for f_1 it settles to a positive value: near the jump discontinuities there are roughly 8% overshoots which get narrower but not smaller. Gibbs phenomenon notwithstanding, we conclude that f_1 is represented by its Fourier series. In fact, if we excise arbitrarily small neighborhoods of jump discontinuities, the Fourier series of f_1 converges uniformly on the remainder of the interval.

We introduced Fourier series in complex form because it is more convenient for computations. For completeness, we will state the real form as well. A real-valued function f on $[0, T]$ can be expanded into the sum of real harmonics

$$a_0 + \sum_{n=1}^{\infty} a_n \cos(n\omega t) + b_n \sin(n\omega t). \tag{9.10}$$

The constant of the series is the average of the function on $[0, T]$

$$a_0 = \frac{1}{T} \int_0^T f(t)\,dt, \tag{9.11}$$

while the remaining coefficients are given by

$$a_n = \frac{2}{T} \int_0^T \cos(n\omega t)\,f(t)\,dt, \quad b_n = \frac{2}{T} \int_0^T \sin(n\omega t)\,f(t)\,dt. \tag{9.12}$$

If the function has jump discontinuities the series (9.10) converges to Fourier value.

9.2 Inner Product Spaces

The dot product is often introduced in early Physics (for the sake of defining work) by the formula

$$\mathbf{x} \cdot \mathbf{y} = \|\mathbf{x}\|\,\|\mathbf{y}\|\cos(\theta). \tag{9.13}$$

This suggests that, in order to compute the dot product, one has to know the lengths of the vectors and the angle between them. In actuality, it is the lengths and the angle that are defined using the dot product:

$$\|\mathbf{x}\| = \sqrt{\mathbf{x}\cdot\mathbf{x}}, \quad \|\mathbf{y}\| = \sqrt{\mathbf{y}\cdot\mathbf{y}}, \quad \cos(\theta) = \frac{\mathbf{x}\cdot\mathbf{y}}{\|\mathbf{x}\|\,\|\mathbf{y}\|}. \tag{9.14}$$

Thus it is the dot product that defines geometry in $\mathbb{R}^n$, rather than the other way around.

The dot product is a special case of an *inner product*. On a real vector space V, an inner product $\langle \cdot, \cdot \rangle$ is a mapping that takes pairs of vectors into real numbers subject to the following axioms:

9.2 Inner Product Spaces

(IP1) Bilinearity: $\langle \sum_i a_i \mathbf{x}_i, \sum_j b_j \mathbf{y}_j \rangle = \sum_i \sum_j a_i b_j \langle \mathbf{x}_i, \mathbf{y}_j \rangle$.
(IP2) Symmetry: $\langle \mathbf{x}, \mathbf{y} \rangle = \langle \mathbf{y}, \mathbf{x} \rangle$.
(IP3) Positivity: $\langle \mathbf{x}, \mathbf{x} \rangle \geq 0$; moreover, $\langle \mathbf{x}, \mathbf{x} \rangle = 0$ implies $\mathbf{x} = 0$.

A bilinear mapping taking pairs of vectors into scalars is called a bilinear *form*; the dot product is one of infinitely many bilinear, symmetric, positive forms on $\mathbb{R}^N$.

A real vector space with an inner product—*an inner product space*—has geometry defined via

$$\|\mathbf{x}\| = \sqrt{\langle \mathbf{x}, \mathbf{x} \rangle}, \quad \cos(\theta) = \frac{\langle \mathbf{x}, \mathbf{y} \rangle}{\|\mathbf{x}\| \|\mathbf{y}\|}. \tag{9.15}$$

Since $\langle \cdot, \cdot \rangle$ is positive definite, the length (2-norm) $\|\mathbf{x}\|$ is guaranteed to be a nonnegative real number and only the zero vector has zero length, which makes intuitive sense.

To show that the definition of the angle in (9.15) is sound, consider the function

$$f(t) = \langle \mathbf{x} - t\mathbf{y}, \mathbf{x} - t\mathbf{y} \rangle = \|\mathbf{x}\|^2 - 2 \langle \mathbf{x}, \mathbf{y} \rangle t + \|\mathbf{y}\|^2 t^2,$$

where $\mathbf{y} \neq 0$. By positivity of $\langle \cdot, \cdot \rangle$, $f(t) \geq 0$. Also, being a quadratic with positive leading coefficient the function f has a global minimum at $t_* = \langle \mathbf{x}, \mathbf{y} \rangle / \|\mathbf{y}\|^2$. Since $f(t)$ is nonnegative, so is its global minimum:

$$f(t_*) = \|\mathbf{x}\|^2 - \frac{\langle \mathbf{x}, \mathbf{y} \rangle^2}{\|\mathbf{y}\|^2} \geq 0.$$

This implies that $|\langle \mathbf{x}, \mathbf{y} \rangle| \leq \|\mathbf{x}\| \|\mathbf{y}\|$, and, therefore

$$-1 \leq \frac{\langle \mathbf{x}, \mathbf{y} \rangle}{\|\mathbf{x}\| \|\mathbf{y}\|} \leq 1.$$

Hence we can interpret the ratio $\langle \mathbf{x}, \mathbf{y} \rangle / \|\mathbf{x}\| \|\mathbf{y}\|$ as cosine of the angle between $\mathbf{x}$ and $\mathbf{y}$. Notice how in this short proof we used all three axioms of the inner product.

For complex vector spaces the inner product is complex-valued and the symmetry requirement (IP2) is replaced with conjugate, or *hermitian symmetry*: $\langle \mathbf{x}, \mathbf{y} \rangle = \overline{\langle \mathbf{y}, \mathbf{x} \rangle}$. The reason for conjugate symmetry is the definition of length, which is the same as in the real case, and the revised definition of the angle between vectors:

$$\|x\| = \sqrt{\langle \mathbf{x}, \mathbf{x} \rangle}, \quad \cos(\theta) = \frac{\mathrm{Re}\left(\langle \mathbf{x}, \mathbf{y} \rangle\right)}{\|\mathbf{x}\| \|\mathbf{y}\|}. \tag{9.16}$$

If $\mathbf{x}$ and $\mathbf{y}$ are interchanged in (9.16), the angle θ does not change. Therefore we must have

$$\mathrm{Re}\left(\langle \mathbf{x}, \mathbf{y} \rangle\right) = \mathrm{Re}\left(\langle \mathbf{y}, \mathbf{x} \rangle\right).$$

Now expand $\langle \mathbf{x} - \mathbf{y}, \mathbf{x} - \mathbf{y} \rangle$ using bilinearity and rearrange the result into

$$\langle \mathbf{x}, \mathbf{y} \rangle + \langle \mathbf{y}, \mathbf{x} \rangle = \langle \mathbf{x}, \mathbf{x} \rangle + \langle \mathbf{y}, \mathbf{y} \rangle - \langle \mathbf{x} - \mathbf{y}, \mathbf{x} - \mathbf{y} \rangle.$$

The right side involves squares of norms and is a real number. The left side is, therefore, also a real number. Consequently,

$$\text{Im}(\langle \mathbf{x}, \mathbf{y} \rangle) = -\text{Im}(\langle \mathbf{y}, \mathbf{x} \rangle).$$

This proves that $\langle \mathbf{x}, \mathbf{y} \rangle$ is the complex conjugate of $\langle \mathbf{y}, \mathbf{x} \rangle$.

As an example of an inner product on $\mathbb{R}^N$ that is different from the dot product, consider

$$\langle x, y \rangle = \sum_{k=1}^{N} w_k \, x_k \, y_k \tag{9.17}$$

where the *weights* w_k are all positive. Weighted inner products are common in regression analysis: the weights may account for different levels of uncertainty or different scales in different measurements.

As another example, which will look familiar, let V be the space of functions on $[0, T]$. Define an inner product on V by

$$\langle f, g \rangle = \int_0^T f(t) \, g(t) \, w(t) \, dt, \quad w(t) > 0. \tag{9.18}$$

This is the weighted L^2 *inner product*, the continuous analog of the weighted dot product (9.17). If $w(t) = 1$ then (9.18) is called the standard L^2 inner product.

Equation (9.6) is the standard complex L^2 inner product on $[0, T]$. In Sect. 9.1 we tacitly assumed that the integral in (9.6) is always finite, yet for some functions that may not be the case. If a function has a bad singularity on $[0, T]$, it may have infinite L^2 norm. This problem is specific to infinite-dimensional spaces: in finite-dimensional spaces the norms of all vectors are finite. In order to ensure that the inner product (9.6) always exists, the space of functions must be restricted by requiring L^2 norms to be finite. This defines what is called the (complex) L^2 space on $[0, T]$:

$$L^2([0, T]) = \{ f : [0, T] \to \mathbb{C} \mid \langle f, f \rangle < \infty \}.$$

The complex exponentials $\phi_n(t) = e^{in\omega t}$ form an orthogonal basis of $L^2([0, T])$. The Fourier series (9.3) is thus an orthogonal basis expansion

$$f(t) = \sum_{n=-\infty}^{\infty} \frac{\langle \phi_n, f \rangle}{\langle \phi_n, \phi_n \rangle} \phi_n(t), \tag{9.19}$$

an infinite-dimensional analogue of (8.2).

9.3 Fourier Series Solutions of ODE

In Sect. 8.6 we used DFT to approximate the solution of a first order scalar linear ODE with constant coefficients (Eq. (8.14)). We will now derive the exact solution of that ODE as a Fourier series. Replacing the right-hand side of (8.14) with its Fourier expansion on $[0, T]$ gives

$$\frac{dx}{dt} + kx = \sum_{n=-\infty}^{\infty} \widehat{f}_n e^{in\omega t}. \tag{9.20}$$

As follows from the principle of superposition, a particular solution of (9.20) is

$$x_p = \sum_{n=-\infty}^{\infty} \frac{\widehat{f}_n}{k + in\omega} e^{in\omega t}, \tag{9.21}$$

so the general solution is $x = C e^{-kt} + x_p(t)$. For a generic initial condition, the constant of integration is $C = x_0 - x_p(0)$.

Next, let us consider a forced 1DOF mass-spring system from Sect. 7.6:

$$m\ddot{x} + r\dot{x} + kx = f, \quad x(0) = \dot{x}(0) = 0, \quad \left(\dot{x} = \frac{dx}{dt}\right). \tag{9.22}$$

Previously, we solved this ODE using multidimensional convolution. Now we expand the right-hand side into a Fourier series on $[0, T]$ and, using the principle of superposition, arrive at the particular solution

$$x_p = \sum_{n=-\infty}^{\infty} \frac{\widehat{f}_n}{m(in\omega)^2 + r(in\omega) + k} e^{in\omega t}. \tag{9.23}$$

The general solution of (9.22) is, therefore, $x = C_1 e^{\lambda_1 t} + C_2 e^{\lambda_2 t} + x_p(t)$. For zero initial conditions the constants C_1 and C_2 are found from the system of equations

$$C_1 + C_2 + x_p(0) = 0, \quad \lambda_1 C_1 + \lambda_2 C_2 + \dot{x}_p(0) = 0.$$

Notice that the denominator in (9.23) is the characteristic polynomial of (9.22) evaluated at $(in\omega)$. Let $p(\lambda)$ be the characteristic polynomial of

$$a_n x^{(n)} + a_{n-1} x^{(n-1)} + a_{n-2} x^{(n-2)} + \cdots + a_0 x = f. \tag{9.24}$$

As follows from the method of undetermined coefficients and the principle of superposition, the Fourier series solution of (9.24) on $[0, T]$ is

$$x_p = \sum_{n=-\infty}^{\infty} \frac{\widehat{f}_n}{p(in\omega)} e^{in\omega t}. \tag{9.25}$$

In (9.25) we assume (as we tacitly did in (9.23)) that none of the numbers $i\,n\,\omega$ coincide with the roots of p; if some do, the corresponding terms in the series need to be handled separately, in the same way that Eq. (7.23) was handled in Sect. 7.4. The general solution of (9.24) is $x = \sum_{k=1}^{n} C_k e^{\lambda_k t} + x_p(t)$ with constants that are found by solving the system

$$\sum_{k=1}^{n} \lambda_k^j C_k + x_p^{(j)}(0) = x^{(j)}(0), \quad j = 0, \ldots, n-1.$$

Moving on, consider the general first order matrix-vector system

$$\frac{d\mathbf{x}}{dt} = A\mathbf{x} + \mathbf{f}, \quad \mathbf{x}(0) = \mathbf{x}_0. \tag{9.26}$$

Expanding all components of $\mathbf{f}$ into Fourier series on $[0, T]$ changes (9.26) to

$$\frac{d\mathbf{x}}{dt} = A\mathbf{x} + \sum_{n=-\infty}^{\infty} \widehat{\mathbf{f}}_n e^{i n \omega t}.$$

Using the method of undetermined coefficients and the principle of superposition, in matrix-vector form, gives the particular solution

$$\mathbf{x}_p = \sum_{n=-\infty}^{\infty} (i n \omega I - A)^{-1} \widehat{\mathbf{f}}_n e^{i n \omega t}, \tag{9.27}$$

where I is the identity matrix of the same size as A. The general solution of (9.26) is then $\mathbf{x} = e^{tA} \mathbf{c} + \mathbf{x}_p(t)$ with the constant $\mathbf{c}$ that should be set to $\mathbf{x}_0 - \mathbf{x}_p(0)$ for generic initial condition.

Finally, let us consider a second order matrix-vector system which may describe a linear mass-spring system with multiple degrees of freedom:

$$M\ddot{\mathbf{x}} + R\dot{\mathbf{x}} + K\mathbf{x} = \mathbf{f}. \tag{9.28}$$

In (9.28) M is the *mass matrix*, R is the *resistance matrix*, and K is the *stiffness matrix*. This ODE can be converted into a first order ODE (9.26) whose solution we have just constructed. However, if just the particular solution is desired, it can be found directly through Fourier expansion of the forcing term $\mathbf{f}$ followed by the same computation as for (9.26):

$$\mathbf{x}_p = \sum_{n=-\infty}^{\infty} \left((i n \omega)^2 M + (i n \omega) R + K\right)^{-1} \widehat{\mathbf{f}}_n e^{i n \omega t}. \tag{9.29}$$

Theoretically, some of the matrices in (9.27) and (9.29) may be singular. However, that simply means that the corresponding terms should be computed as limits of indeterminate forms.

9.4 RC-Circuit Driven by a Square Waveform

Let $a, S > 0$. On $[0, S]$ a square waveform of amplitude a and period S is defined by

$$f(t) = \begin{cases} a, & 0 < t < S/2, \\ -a, & S/2 < t < S, \\ 0, & t = 0, S/2, S; \end{cases} \quad (9.30)$$

outside $[0, S]$ it is defined by S-periodic extension: $f(t + S) = f(t)$. The subject of this section is the voltage V across the capacitor in an RC-circuit driven by (9.30). Theoretically, it satisfies the IVP

$$\frac{dV}{dt} + \frac{1}{RC} V = f, \quad V(0) = V_0. \quad (9.31)$$

In Chap. 6 (Sect. 6.8) ODE (9.31) was used to illustrate the principle of superposition and offer a glimpse of Fourier series. We will now derive the Fourier series solution of (9.31) from first principles.

Let $\omega = 2\pi/S$. The Fourier series coefficients of (9.30) are

$$\widehat{f_n} = \frac{a}{S} \left\{ \int_0^{\frac{S}{2}} e^{-in\omega t} dt - \int_{\frac{S}{2}}^{S} e^{-in\omega t} dt \right\} = -ai \frac{(1 - e^{i\omega n S/2})^2}{\omega n S}$$

$$= -\frac{ai}{2} \frac{(1 - e^{i\pi n})^2}{\pi n} = \begin{cases} 0, & \text{if } n \text{ is even,} \\ -2ai/(\pi n), & \text{if } n \text{ is odd.} \end{cases} \quad (9.32)$$

As follows from the discussion in Sect. 9.3, the Fourier series solution of (9.31) is the sum of the particular solution

$$V_p = \sum_{n=-\infty}^{\infty} \frac{\widehat{f_n}}{1 + RCi\omega n} e^{in\omega t} \quad (9.33)$$

with $\widehat{f_n}$ given by (9.32) and the complimentary function $V_c = C e^{-t/RC}$; the constant of integration C should be set to $V_0 - V_p(0)$.

Fourier series (9.33) has a closed form expression which can be deduced as follows. On $[0, S/2]$ Eq. (9.33) is a solution of (9.31) with right-hand side $f = a$, which is

$$V_p = a + (V_p(0) - a) e^{-t/RC}, \quad 0 \leq t \leq S/2;$$

on $[S/2, S]$ it is a solution of (9.31) with right-hand side $f = -a$, which is

$$V_p = -a + a (V_p(S/2) + a) e^{-(t-S/2)/RC}, \quad S/2 \leq t \leq S.$$

Since V_p is continuous on $[0, S]$

$$V_p(S/2) = a + (V_p(0) - a) e^{-S/2RC}, \quad (9.34)$$

and, since it is periodic,

$$V_p(0) = V_p(S) = -a + a\left(V_p(S/2) + a\right) e^{-S/2RC}. \tag{9.35}$$

From Eqs. (9.34) and (9.35) follows that

$$V_p(0) = V_p(S) = -a \frac{1 - e^{-S/2RC}}{1 + e^{-S/2RC}}.$$

Consequently, for $t \in [0, S]$ the sum of the Fourier series (9.33) is, exactly,

$$V_p = a \begin{cases} 1 - c e^{-\frac{t}{RC}}, & 0 < t < S/2, \\ -1 + c e^{-\frac{t-S/2}{RC}}, & S/2 < t < S, \end{cases} \quad c = \frac{2}{1 + e^{-S/2RC}}; \tag{9.36}$$

outside $[0, S]$ V_p is defined by periodic extension.

The bottom panel of Fig. 9.2 shows the Fourier series solution of (9.31) along with the exact solution and the output of ode45; the top panel shows the right-hand side of (9.31) and its Fourier approximation. We set $RC = 22$ ms, $S = 1$ s, $a = 5$ V, $V_0 = 5$ V, and used 10 complex harmonics in the Fourier approximations; the parameters were chosen close to the values used in an experiment considered later.

To produce Fig. 9.2, we first declared subfunctions for computing exact values of (9.30) and (9.33).

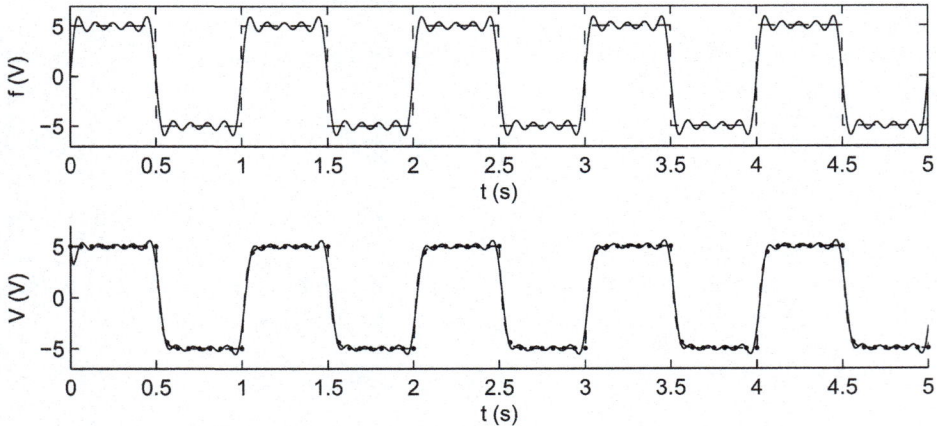

Fig. 9.2 Solution of (9.31) with $RC = 22$ ms, $S = 1$ s, $a = 5$ V, $V_0 = 5$ V. The top panel shows the right-hand side of (9.31) (dashed line) and its Fourier series approximation (solid line); the bottom panel shows the exact solution of (9.30) (dashed line), its Fourier approximation with 10 complex harmonics (solid line), and numerical solution computed using ode45 (dots)

9.4 RC-Circuit Driven by a Square Waveform

```
RC = 22e-3; S = 1; a = 5;

function y = square_wave(t)
t = mod(t,S); y = zeros(size(t));
y(t < .5*S) =  a; y(t > .5*S) = -a;
end

function vp = particular(t)
t = mod(t,S); vp = zeros(size(t));
c = 2/(1 + exp(-.5*S/RC));
ind = (t <= .5*S);
vp(ind) =  a*(1 - c*exp(-t(ind)/RC));
ind = (t > .5*S);
vp(ind) =  a*(-1 + c*exp(-(t(ind)-.5*S)/RC));
end
```

Then the numerical solution was computed with ode45 on an equispaced grid covering the interval [0, 16]. To ensure good accuracy, relative tolerance was set to 1×10^{-6}.

```
odefun = @(t,v) (square_wave(t) - v)/RC; v0 = a;
T = 16; N = 8192; t = T*(0:N-1)/N;
RelTol = 1e-6; opts.RelTol = RelTol;
[t,v] = ode45(odefun,t,v0,opts);
```

The rest of the code is as follows. Since MATLAB is an interpreted language, it is usually advisable to replace loops with matrix multiplication, whenever possible. We did not follow that practice when computing Fourier approximations both for clarity, and because it does not speed up the code.

```
v(1,3) = 0; v(:,2) = particular(t);
v(:,2) = v(:,2) + (v0 - v(1,2))*exp(-t/RC);
y = square_wave(t); y(1,2) = 0;

w = 2*pi/S; b = 9;
for n=-b:2:b
    fhat = -2*a*1i/pi/n;
    y(:,2) = y(:,2) + fhat*exp(1i*n*w*t);
    fhat = fhat/(1 + RC*1i*n*w);
    v(:,3) = v(:,3) + fhat*exp(1i*n*w*t);
end

figure;
subplot(2,1,1);
plot(t,y(:,1),'k--',t, real(y(:,2)),'k-');
```

```
xlim([0 5*S]);
ylim((a+2)*[-1 1]);
xlabel('t (s)'); ylabel('f (V)');
subplot(2,1,2); hold on;
plot(t,v(:,2),'k--');
plot(t,real(v(:,3)),'k-');
plot(t(1:32:N),v(1:32:N,1),'k.');
xlim([0 5*S]);
ylim((a+2)*[-1 1]);
xlabel('t (s)'); ylabel('V (V)');
```

Returning to Fig. 9.2, It is clear that, even for 10 harmonics, the Fourier series approximation of the solution of (9.31) is in good agreement with the exact solution. Increasing the bandwidth b improves the agreement and, since (9.33) is continuous but not differentiable, the convergence is linear.

It is instructive to perform DFT analysis of the data in Fig. 9.2. The top panel of Fig. 9.3 shows the frequency content of the exact square wave; the bottom panel shows that of the exact solution of (9.31). In order to generate Fig. 9.3 we applied fft as described in Chap. 8.

```
ww = (0:N-1)/T;
yy = fft(y(:,1))/N;   vv = fft(v(:,2))/N;

figure;
subplot(2,1,1); plot(ww,2*abs(yy),'k-');
xlim([0 30]); xlabel('w (Hz)'); ylabel('2 |fhat| (V)');
subplot(2,1,2); plot(ww,2*abs(vv),'k-')
xlim([0 30]); xlabel('w (Hz)'); ylabel('2 |Vhat| (V)');
```

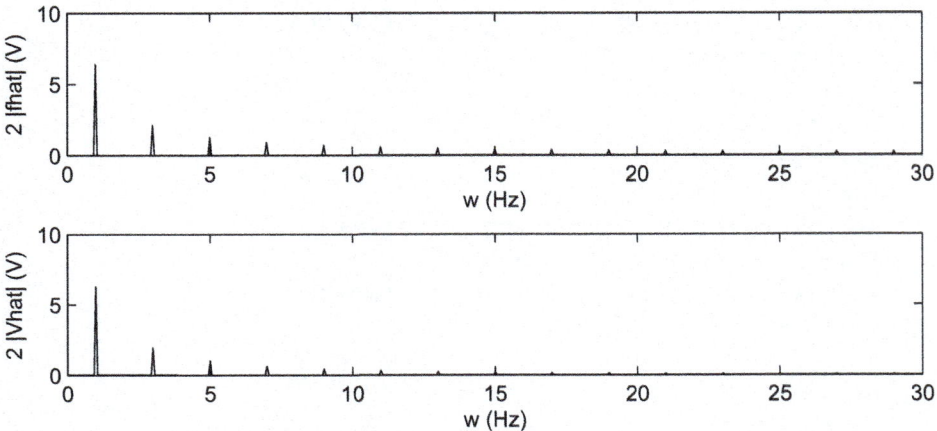

Fig. 9.3 DFT analysis of the data in Fig. 9.2

9.4 RC-Circuit Driven by a Square Waveform

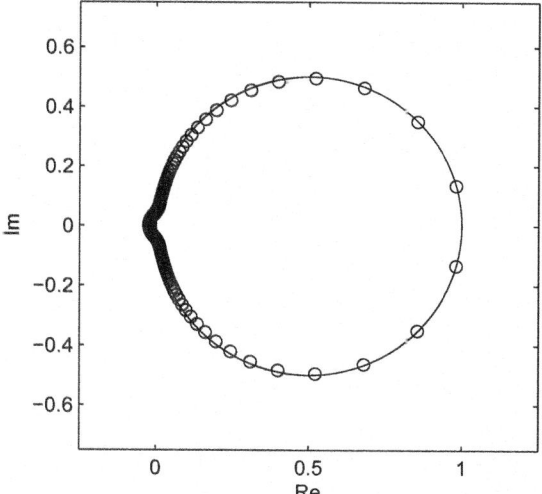

Fig. 9.4 Complex plot of (9.37) extracted from the DFT data in Fig. 9.3

Since the square wave has unit period, its spectrum consists of odd integer frequencies, in accordance with (9.32). For the parameters used in the computation, the contribution from the complimentary function is small and the solution of (9.31) is close to (9.33): that is why the two spectra in Fig. 9.3 look so similar.

The following code extracts the DFT coefficients corresponding to the peaks and computes their ratios, which closely approximate the numbers

$$\frac{1}{1 + RCi\omega n}, \quad n = 0, \ldots, N/2. \tag{9.37}$$

When plotted in the complex plane along with their conjugates, these complex numbers land on the circle shown in Fig. 9.4.

```
ind = T+1:2*T:.5*N;
h = vv(ind)./yy(ind);
theta = linspace(0,2*pi);

figure; axis equal;
plot([h conj(h)],'ko'); hold on;
plot(.5 + .5*cos(theta),.5*sin(theta),'k-');
xlabel('Re'); ylabel('Im');
xlim([-.25 1.25]); ylim([-.75 .75]);
```

Indeed, the numbers (9.37) lie on the curve parameterized by

$$x(s) = \text{Re}\left(\frac{1}{1+is}\right) = \frac{1}{1+s^2}, \quad y(s) = \text{Im}\left(\frac{1}{1+is}\right) = \frac{-s}{1+s^2}, \quad -\infty < s < \infty.$$

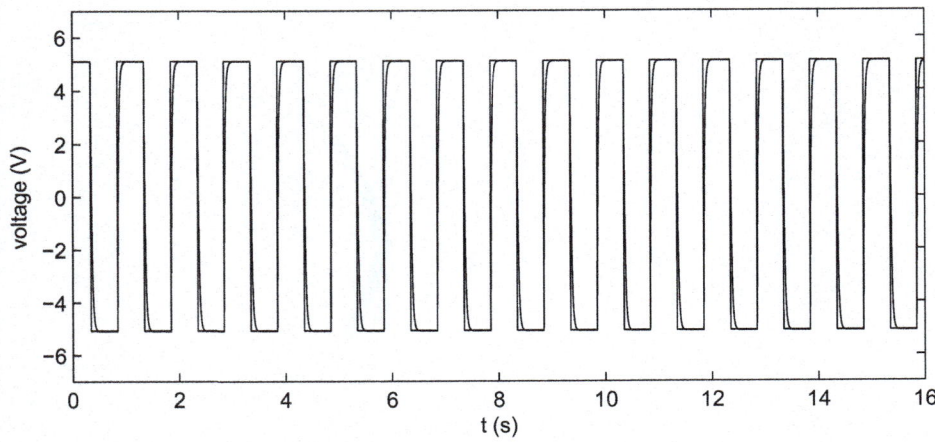

Fig. 9.5 RC-circuit driven by a square waveform; shown is the voltage across the capacitor and the square waveform itself

Since

$$\left(x - \frac{1}{2}\right)^2 + y^2 = \frac{1}{4}$$

that curve is a circle of radius $1/2$ centered at $(1/2, 0)$.

The alignment in Fig. 9.4 is not perfect because of the errors in the DFT. These errors tend to increase with frequency which results in a bulge around the origin.

To see how well theory matches reality, we collected the data shown in Fig. 9.5 for an RC-circuit from Sect. 3.4 using the same instrumentation and methodology. The circuit has a $2.2\,\mu\text{F}$ capacitor and a $10\,\text{k}\Omega$ resistor which correspond to the nominal value $RC = 22$ ms; as we found in Sect. 3.4, the actual value of the RC-constant is about 22.8 ms. We drove the circuit with a square waveform having period $S = 1$s and amplitude $V = 5$ V. The data was sampled at the rate of 512 Hz which for 8192 measurements corresponds to 16 s of observations. The voltage across the capacitor looks a lot like the square waveform because the latter's period is large compared to the RC-constant. Decreasing S would make the plot in Fig. 9.5 visually more interesting, but we need the fundamental frequency of the square waveform to be as small as possible.

Figure 9.6 shows the frequency content of experimental data. Notice how close that is to the frequency content of simulated data in Fig. 9.3.

The plot in Fig. 9.7 is closer to the theoretical circle than the one in Fig. 9.4. One explanation for that is the lesser influence of the complimentary function. The data was collected after the circuit had been run for some time, enough for the complimentary function to completely die out.

The numbers (9.37) can also be used to estimate the RC-constant. A quick estimate using `fminsearch` gave 22.6 s which is close to the value obtained in Sect. 3.4. It is not exactly the same because of the errors in the DFT.

9.5 Heat Equation on an Interval

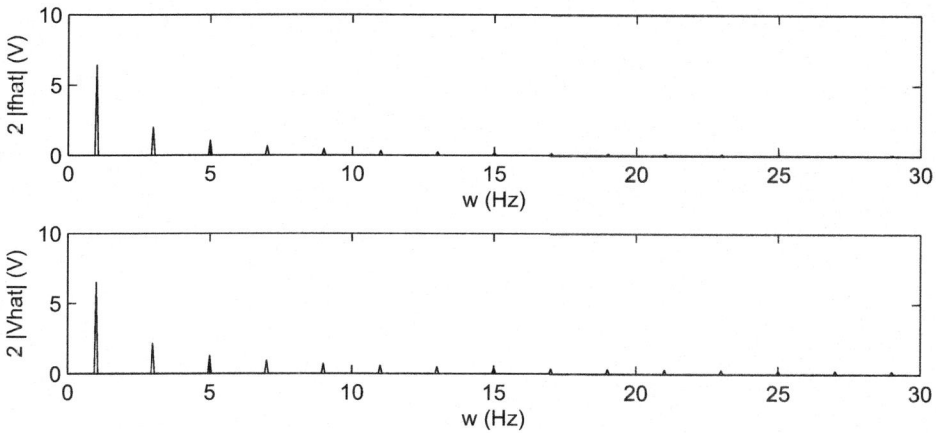

Fig. 9.6 DFT analysis of the data in Fig. 9.5. The top panel shows the frequency content of the square wave; the bottom panel shows that of the capacitor voltage

Fig. 9.7 Complex plot of (9.37) computed from the DFT data in Fig. 9.6

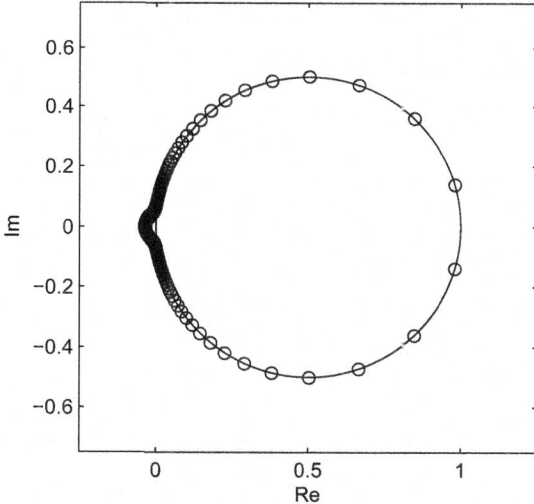

9.5 Heat Equation on an Interval

Historically, Fourier series first appeared as solutions of partial differential equations (PDE). This is an appropriate place to introduce a simple PDE and show what its Fourier series solution looks like.

Imagine a metal rod in the shape of a circular cylinder having length L, cross-sectional area A, density ρ, and heat capacitance c. Suppose that the lateral surface of the rod is perfectly insulated and that the initial temperature varies only along the axis: this implies that the temperature of the rod varies only along the axis at all times. We can identify the rod with the interval $[0, L]$ and denote its temperature $u(x, t)$ with $0 \leq x \leq L$ and $t \geq 0$. The PDE of the title describes the evolution of $u(x, t)$ from the initial value $u(x, 0) = f(x)$.

The derivation of the heat equation begins with energy balance. Consider an arbitrary subinterval $[a, b]$ contained in $[0, L]$. The total amount of thermal energy Q contained in that subinterval is

$$Q = \int_a^b c\, u(x, t)\, \rho\, A\, dx.$$

Not much can be said about Q itself, so we turn our attention to its rate of change:

$$\frac{dQ}{dt} = \frac{d}{dt} \int_a^b c\, u(x, t)\, \rho\, A\, dx. \tag{9.38}$$

Since the lateral surface of the rod is insulated, the energy content of $[a, b]$ changes only due to heat flux across the boundary cross-sections at $x = a$ and $x = b$:

$$\frac{dQ}{dt} = -A\, \phi(b, t) + A\, \phi(a, t). \tag{9.39}$$

Combining Eqs. (9.38) and (9.39), and canceling the cross-sectional area A, leads to

$$\frac{d}{dt} \int_a^b c\, u(x, t)\, \rho\, dx = -\phi(b, t) + \phi(a, t). \tag{9.40}$$

We arrived at a global conservation law—energy balance applied to a finite, as opposed to an infinitesimal portion of the rod. In order to pass to a more useful local conservation law, we manipulate (9.40) as follows. On the left-hand side, interchange integration and differentiation—the full derivative becomes a partial derivative once brought inside the integral sign:

$$\frac{d}{dt} \int_a^b c\, u(x, t)\, \rho\, dx = \int_a^b c\, \rho\, \frac{\partial u}{\partial t}(x, t)\, dx.$$

Next, use the Fundamental Theorem of Calculus to rewrite the right-hand side of (9.40) as an integral

$$-\phi(b, t) + \phi(a, t) = -\int_a^b \frac{\partial \phi}{\partial x}(x, t)\, dx$$

and combine it with the left-hand side

$$\int_a^b \left(c\, \rho\, \frac{\partial u}{\partial t}(x, t) + \frac{\partial \phi}{\partial x}(x, t) \right) dx = 0.$$

Due to the arbitrary choice of the section $[a, b]$, the above integral vanishes for all choices of limits. This implies that the integrand must vanish:

9.5 Heat Equation on an Interval

$$c\rho \frac{\partial u}{\partial t}(x,t) + \frac{\partial \phi}{\partial x}(x,t) = 0. \tag{9.41}$$

Equation (9.41) is the sought local conservation law, called *equation of continuity*.

Since (9.41) contains two unknown functions, the temperature $u(x,t)$ and flux $\phi(x,t)$, we need another equation relating these quantities. That equation is Fourier's law of heat conduction, briefly mentioned in Sect. 3.10. According to Fourier, the heat flux is proportional to the gradient of temperature and is directed against it. In one dimension, that translates into $\phi = -k\, \partial u/\partial x$, where k is thermal conductivity.

Combining the equation of continuity (9.41) with Fourier's law of heat conduction gives the heat equation

$$\frac{\partial u}{\partial t} = D \frac{\partial^2 u}{\partial x^2}, \quad \left(D = \frac{k}{c\rho}\right). \tag{9.42}$$

The aggregated constant D is called *thermal diffusivity*.

In addition to an initial condition, Eq. (9.42) needs *boundary conditions* which specify what happens at the boundary of the domain $[0, L]$. The two main types of boundary conditions are the following:

(BC1) Dirichlet : $u(x,t)$ is prescribed at $x = 0, L$.
(BC2) Neumann : $\frac{\partial u}{\partial t}(x,t)$ is prescribed at $x = 0, L$.

Together with a boundary condition the heat equation comprises a *boundary value problem* (BVP).

If the rod is bent into a circular ring, the evolution of temperature is still described by the heat equation (9.42), but the endpoints become one and the same point on the ring, hence $u(0,t) = u(L,t)$: these are *periodic boundary conditions*. The BVP for the temperature of a circular ring is

$$\begin{aligned}\frac{\partial u}{\partial t} &= D\frac{\partial^2 u}{\partial x^2}, \quad 0 < x < L, \quad t > 0, \\ u(0,t) &= u(L,t), \quad u(x,0) = f(x).\end{aligned} \tag{9.43}$$

It is slightly easier to analyze than the Dirichlet and Neumann problems, so we will solve it first.

Since the solution of (9.43) is L-periodic, it can be written as the Fourier series

$$u(x,t) = \sum_{n=-\infty}^{\infty} \widehat{u}_n(t)\, e^{in\omega t}, \quad \omega = \frac{2\pi}{L}. \tag{9.44}$$

Substituting (9.44) into (9.43) gives

$$\sum_{n=-\infty}^{\infty} \frac{d\widehat{u}_n}{dt} e^{in\omega t} = \sum_{n=-\infty}^{\infty} D\widehat{u}_n(t) (in\omega)^2 e^{in\omega t}$$

$$\sum_{n=-\infty}^{\infty} \widehat{u}_n(0) e^{in\omega t} = f(x) = \sum_{n=-\infty}^{\infty} \widehat{f}_n e^{in\omega t}.$$

Equating like Fourier coefficients shows that

$$\frac{d\widehat{u}_n}{dt} = -Dn^2\omega^2 \widehat{u}_n, \quad \widehat{u}_n(0) = \widehat{f}_n$$

and, therefore,

$$\widehat{u}_n = \widehat{f}_n e^{-Dn^2\omega^2 t}. \tag{9.45}$$

Consequently, the solution of (9.43) is

$$u(x,t) = \sum_{n=-\infty}^{\infty} \widehat{f}_n e^{-Dn^2\omega^2 t} e^{in\omega t}. \tag{9.46}$$

Figure 9.8 shows the plots of (9.43) with $L = 2\pi$ and $D = 1$ for several values of time. We chose the initial condition with jump discontinuities to show how quickly those discontinuities get smoothed out.

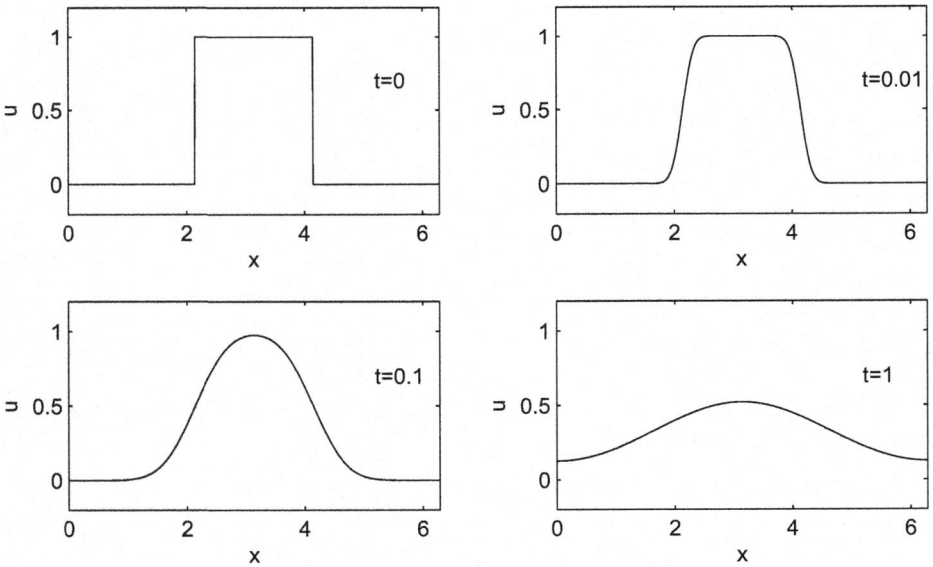

Fig. 9.8 Solution of (9.43) with $L = 2\pi$ and $D = 1$. The summation in the Fourier series (9.46) was carried out for $|n| \leq 50$. The initial condition is shown in the upper left panel

9.5 Heat Equation on an Interval

The figure also shows that, as $t \to \infty$, the temperature of the ring becomes constant. Indeed, according to (9.44), for $n \neq 0$ $\lim_{t \to \infty} \widehat{u}_n = 0$. Therefore

$$\lim_{t \to \infty} u(x,t) = \widehat{u}_0 = \widehat{f}_0 = \frac{1}{L} \int_0^L f(x)\,dx.$$

The temperature of a perfectly insulated ring tends to the average of the initial temperature.

For non-periodic boundary conditions one uses Fourier series consisting of functions that satisfy those conditions. These functions are found using separation of variables—technically, a different method from the one discussed in Sect. 1.3, but based on the same principle.

As an example of separation of variables in the context of PDE, consider the Neumann problem

$$\frac{\partial u}{\partial t} = D \frac{\partial^2 u}{\partial x^2}, \quad 0 < x < L, \quad t > 0,$$
$$\frac{\partial u}{\partial x}(0,t) = \frac{\partial u}{\partial x}(L,t), \quad u(x,0) = f(x). \tag{9.47}$$

The Fourier series solution of (9.43), given by Eq. (9.46), is the sum of terms of the form

$$(\text{function of } t) \times (\text{function of } x).$$

Let us assume that the same holds for the Fourier series solution of (9.47). Setting $u(x,t) = \psi(t)\phi(x)$ in the heat equation results in

$$\frac{d\psi}{dt}\phi(x) = D\,\psi(t)\frac{d^2\phi}{dx^2}.$$

"Separating variables" now means collecting all functions of t on one side, and all functions of x on the other:

$$\frac{1}{D\,\psi(t)}\frac{d\psi}{dt} = \frac{1}{\phi(x)}\frac{d^2\phi}{dx^2}.$$

Since x and t are independent variables, the only way a function of t can equal a function of x is if both are constant. Therefore,

$$\frac{d\psi}{dt} = D\lambda\psi, \quad \frac{d^2\phi}{dx^2} = \lambda\phi,$$

and, consequently,

$$\psi = C\,e^{D\lambda t}, \quad \phi = A\,e^{\sqrt{\lambda}\,x} + B\,e^{-\sqrt{\lambda}\,x}.$$

Imposing Neumann boundary conditions on ϕ gives the linear homogeneous system

$$\sqrt{\lambda}\,(A - B) = 0, \quad \sqrt{\lambda}\,(A\,e^{\sqrt{\lambda}\,L} - B\,e^{-\sqrt{\lambda}\,L}) = 0.$$

If $\lambda = 0$ then both ϕ and ψ are constant and we get a constant solution of (9.47). If $\lambda \neq 0$ then it can be canceled and from the first equation follows that $A = B$. From the second equation then follows that $e^{\sqrt{\lambda}L} - e^{-\sqrt{\lambda}L} = 0$, otherwise $\phi \equiv 0$. We thus get a restriction on λ which can be rewritten as $e^{2\sqrt{\lambda}L} = 1$ so that it can be compared with Euler's formula. As follows from that comparison, $2\sqrt{\lambda}L = 2\pi i n$ where n is an integer. It follows that (9.47) has separated solutions of the form

$$C e^{-D\pi^2 n^2 t/L} \left(A e^{\pi i n x/L} + A e^{-\pi i n x/L} \right) \sim e^{-D\pi^2 n^2 t/L} \cos(\pi n x/L).$$

By the principle of superposition, same one as in Sect. 6.8, the series

$$\sum_{n=0}^{\infty} c_n e^{-D\pi^2 n^2 t/L} \cos(\pi n x/L) \qquad (9.48)$$

is a solution of the heat equation that satisfies Neumann boundary conditions. For it to be the solution of (9.47) the coefficients c_n should be chosen to satisfy the initial condition:

$$\sum_{n=0}^{\infty} c_n \cos(\pi n x/L) = f(x).$$

It so happens that the cosines $\cos(\pi n x/L)$ are orthogonal on $[0, L]$ with respect to the standard L^2 inner product:

$$\langle \cos(\pi n x/L), \cos(\pi m x/L) \rangle = \begin{cases} 0, & n \neq m, \\ L, & n = m = 0, \\ L/2, & n = m \neq 0. \end{cases}$$

As follows from orthogonality,

$$\begin{aligned} c_0 &= \frac{1}{L} \int_0^L f(x)\, dx, \\ c_n &= \frac{2}{L} \int_0^L f(x) \cos(\pi n x/L)\, dx, \quad n > 0. \end{aligned} \qquad (9.49)$$

We conclude that the solution of (9.47) is the Fourier series (9.48) with coefficients given by (9.49).

9.6 Comments and Bibliography

Trigonometric series first appeared in the works of Euler, d'Alembert, Daniel Bernoulli, and Gauss. Yet it was Fourier who first revealed their true power in his 1807 manuscript on propagation of heat in solids. In 1811 the Paris Institute offered a prize for a mathematical

9.6 Comments and Bibliography

theory of heat conduction. Fourier submitted his 1807 manuscript and won the prize, but not without controversy. A committee, consisting of Lagrange, Laplace, Legendre, Haüy, and Malus, stated in its report that

> ...the manner in which the author arrives at these equations is not exempt of difficulties and that his analysis to integrate them still leaves something to be desired on the score of generality and even rigour.

The committee had the right to be sceptical because Fourier did not provide any formal proofs and took convergence of the series, which now bear his name, for granted.

The study of convergence of Fourier series was initiated by Dirichlet around 1829 and continued for well over a century. Dirichlet proved that piecewise continuous functions converge to their Fourier values. By "convergence" he meant the existence of the limit: $\lim_{N\to\infty} \sum_{n=-N}^{N} c_n e^{in\omega t}$. There are other ways of defining convergence of Fourier series: using summability kernels, or as convergence in norm. Accordingly, there are many convergence theorems for Fourier series, some of which can be found in [1] and [2]. The last major results concerning convergence of one-dimensional Fourier series were established by Carleson in 1965; many questions about multidimensional Fourier series are still open.

Pointwise convergence of Fourier series is a particularly thorny subject and we must confess that our presentation of it was, out of necessity, almost as lax as Fourier's. One reason why Fourier series are much more difficult than, say, Taylor series, is that they can represent amazingly complicated functions. Figure 9.9 shows the plots of

$$w(t) = \sum_{n=1}^{\infty} \frac{\sin(\pi\, 3^n t)}{\pi\, 2^n} \tag{9.50}$$

on increasingly smaller subintervals of $[0, 1]$. No matter how small the subinterval, the graph of (9.50) looks like the trajectory of a particle executing Brownian motion.

The Fourier series (9.50) is absolutely convergent and the function w it defines is continuous, yet it is nowhere differentiable; it is an example of a *lacunary series* which were discovered by Weierstrass. By the way, Brownian motion can be studied using Fourier analysis. This was demonstrated by Norbert Wiener in 1926.

Inner product spaces were defined and rigorously studied by Hilbert who wanted to find an infinite-dimensional analog of Euclidean space. Hilbert showed that, in order for an infinite-dimensional inner product space V to be "just like $\mathbb{R}^N$", it has to be complete: roughly speaking, every sequence of vectors that ought to converge must converge to an element of the space. In honor of Hilbert, complete inner product spaces are now called *Hilbert spaces*; $L^2([0, T])$ is a prominent example.

Fourier series representation of solutions of linear ODE with constant coefficients is not merely an alternative to the methods of Chap. 7, it is a frame of mind. Consider the simple IVP

$$\frac{dx}{dt} + kx = f, \quad x(0).$$

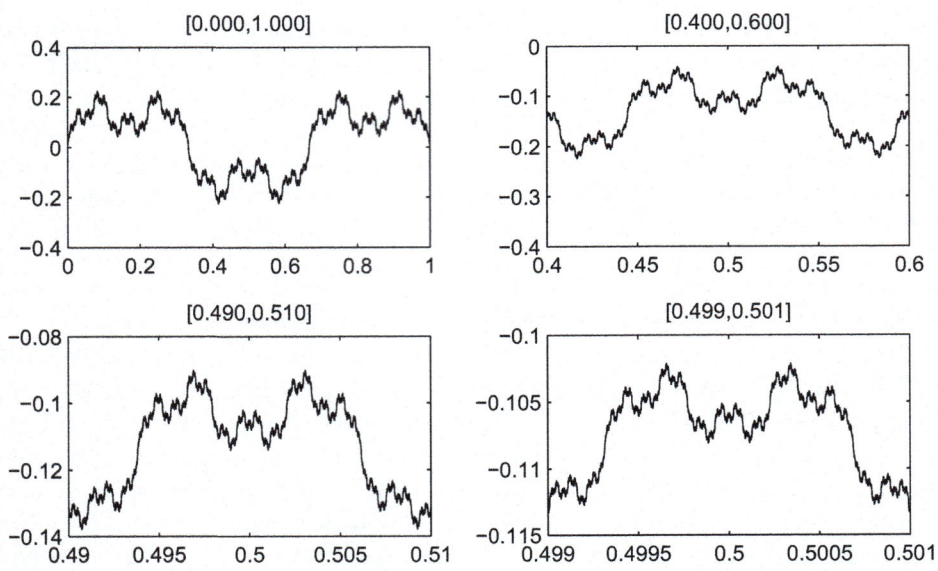

Fig. 9.9 The plot of (9.50)

Convolution $x = e^{-kt} * f$ is a complicated relation between the forcing term f and the solution x. Meanwhile, the relationship between Fourier coefficients $\widehat{f_n}$ and $\widehat{x_n}$ is a simple proportionality

$$\widehat{x_n} = \frac{1}{k + i n \omega} \widehat{f_n}.$$

As we showed in Sect. 9.4, the numbers $h(i\,n\,\omega) = \widehat{x_n}/\widehat{f_n}$ can be extracted from experimental data using DFT and plotted in the complex plane to identify a circuit (such plots are usually called *Nyquist plots*). These numbers are actually values of the *transfer function*, a key concept in system analysis, which will be defined in Chap. 10.

In Sect. 3.10 we described how the discovery of the law of heat conduction by Fourier inspired Ohm and Fick to discover analogous laws for electrical conduction and mass diffusion. If in the derivation given in Sect. 9.5 Fourier's law is replaced with Ohm's or Fick's, the resulting analogue of the heat equation describes one-dimensional diffusion of charge or mass, respectively. For that reason the heat equation is often called the *diffusion equation*.

9.7 Exercises

1. Divide [0, 1] into four equal subintervals and define a piecewise function to be a cubic on each subinterval. Choose the coefficients of the four cubics so that the function is twice continuously differentiable on [0, 1]; the third derivative will have jump discontinuities.

9.7 Exercises

Show that the Fourier approximation $\sum_{n=-N}^{N} c_n e^{in\omega t}$ converges cubically by producing a plot like that in Fig. 9.1.

2. The polynomials

$$p_0 = 1, \quad p_1 = x, \quad p_2 = \frac{3}{2}x^2 - \frac{1}{2}, \quad p_3 = \frac{5}{2}x^3 - \frac{3}{2}x$$

are orthogonal with respect to the standard L^2 inner product on $[-1, 1]$:

$$\langle p_i, p_j \rangle = \int_{-1}^{1} p_i(x) \, p_j(x) \, dx = 0, \quad i \neq j.$$

Find p_n for $n = 4, 5, 6, 7$. Then find the best approximation of the exponential function $f(x) = e^{-x}$ with $g(x) = \sum_{n=0}^{7} c_i \, p_i(x)$ by minimizing the square of the norm of the residual $\|f - g\|^2 = \langle f - g, f - g \rangle$ with respect to the coefficients c_i. What does this tell you about Fourier coefficients?

3. Let $a, S > 0$. On $[0, S]$ define $f(t) = a\left((2/S)t - 1\right)$ and extend it periodically to the entire real line: this is the *sawtooth waveform*. Find the Fourier series solution of the mass-spring system

$$m\ddot{x} + r\dot{x} + kx = f, \quad x(0) = \dot{x}(0) = 0 \tag{9.51}$$

and validate it using ode45 using parameters of your choice. Use DFT to compare the spectra of x and f in the manner of Fig. 9.3.

4. In Sect. 9.4 we generated what is called a Nyquist plot for an RC-circuit (which should be a circle), first from simulated data (Fig. 9.4) and then from real data (Fig. 9.7). Use simulated or real data (or both) to generate a similar plot for an RLC-circuit.

5. The linear wave equation is the following PDE:

$$\frac{\partial^2 u}{\partial t^2} = \frac{1}{c^2} \frac{\partial^2 u}{\partial x^2}.$$

It is commonly introduced on an interval $[0, L]$ with Dirichlet boundary conditions $u(0, t) = u(L, t) = 0$ as a model of a vibrating string. Use separation of variables to find the Fourier series solution of

$$\frac{\partial^2 u}{\partial t^2} = \frac{1}{c^2} \frac{\partial^2 u}{\partial x^2}, \quad 0 < x < L, \quad t > 0,$$

$$u(0, t) = u(L, t) = 0, \quad u(x, 0) = f(x), \quad \frac{\partial u}{\partial t}(x, 0) = g(x). \tag{9.52}$$

The parameter c has the units of velocity; the initial conditions represent the initial displacement and velocity of the string. Plot the solution of (9.52) for a few values of time. Use a simple choice of parameters and initial conditions, for instance, set the initial velocity to zero and the initial displacement to the triangular profile

$$f(x) = \begin{cases} x, & 0 \le x \le a, \\ a(L-x)/(L-a), & a \le x \le L \end{cases}$$

corresponding to a simple pluck. Do the plots conform with your intuition about a vibrating string? If static plots do not tell you much, try to make an animation.

References

1. Dunham Jackson. *Fourier Series and Orthogonal Polynomials*
2. S.G. Krantz. *A Panorama of Harmonic Analysis*. The Carus Mathematical Monographs

Fourier and Laplace Transforms

Let

$$f(t) = \begin{cases} t(1-t)^2 \cos(40t), & 0 \le t \le 1, \\ 0, & \text{otherwise.} \end{cases}$$

Figure 10.1 shows the plot of f and the plot of its Fourier series

$$\tilde{f} = \sum_{n=-\infty}^{\infty} \widehat{f_n} e^{in\omega t}. \tag{10.1}$$

for $T = 3, 6, 9, 12$.

Being a T-periodic extension of f, $\tilde{f}$ agrees with f on $[-T+1, T]$: that is why, as T is increased, the plots of f and $\tilde{f}$ become indistinguishable in the bottom panel of Fig. 10.1. This suggests that in the limit, as $T \to \infty$, the Fourier series $\tilde{f}$ should agree with f on the entire real line.

In Sect. 10.1 we identify the limit of a Fourier series as a Fourier integral—the Fourier transform. We then show, through examples, how Fourier transforms are computed and validated. The section is concluded with three properties of the Fourier transform—the shifting property, the transformation of the derivative property, and the convolution property—which are useful later.

In Sect. 10.2 we use the Fourier transform to solve linear ODE with constant coefficients. We further show that all information about solutions of such ODE is contained in *transfer functions* which can be computed as ratios of Fourier transforms.

In Chap. 9 we used Fourier series to solve the heat equation on an interval. In Sect. 10.3 we solve the heat equation on the entire real line. The solution is expressed as convolution of the initial temperature with the *heat kernel*—an important object in analysis.

© The Author(s), under exclusive license to Springer Nature Switzerland AG 2025
A. Beltukov, *Differential Equations and Data Analysis*, Synthesis Lectures on Mathematics & Statistics, https://doi.org/10.1007/978-3-031-62257-1_10

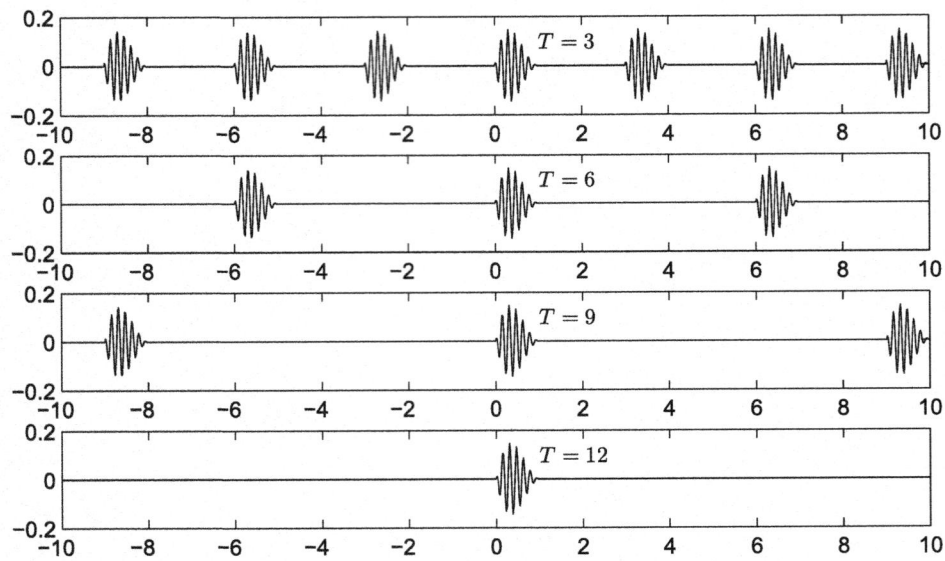

Fig. 10.1 Plots of (10.1) for $T = 3, 6, 9, 12$; $f = t(1-t)^2 \cos(40t)$ on $[0, 1]$

In the final Sect. 10.4 we give homage to Laplace transform which is dominant in system analysis and many other engineering disciplines. The Laplace transform is intimately related to Fourier transform. While it may be slightly more convenient in certain computations, Fourier transform is foundationally more important.

10.1 Fourier Transform

Consider the limit, as $T \to \infty$, of the Fourier series (10.1), where f is now any continuous function vanishing outside $[0, a]$. Figure 10.1 suggests that that limit should equal the function f, which is, indeed, the case. In order to identify the limit, let us first define the *Fourier transform* of f as

$$\widehat{f}(\xi) = \int_{-\infty}^{\infty} f(t) e^{-i\xi t}\, dt, \quad \xi \in \mathbb{R}. \tag{10.2}$$

For $T > a$ the Fourier coefficients $\widehat{f}_n$ in (10.1) may be regarded as sampled values of $\widehat{f}$ divided by $T = 2\pi/\omega$:

$$\widehat{f}_n = \frac{1}{T}\int_0^T f(t) e^{-in\omega t}\, dt = \frac{1}{T}\int_{-\infty}^{\infty} f(t) e^{-in\omega t}\, dt = \frac{\omega}{2\pi} \widehat{f}(n\omega).$$

10.1 Fourier Transform

Accordingly,

$$\lim_{T \to \infty} \tilde{f} = \frac{1}{2\pi} \lim_{\omega \to 0} \sum_{n=-\infty}^{\infty} \widehat{f}(n\omega) e^{i n \omega t} \omega,$$

which shows that we are taking the limit of a Riemann sum corresponding to the integral of $\widehat{f}(\xi) e^{i \xi t}$:

$$f(t) = \frac{1}{2\pi} \int_{-\infty}^{\infty} \widehat{f}(\xi) e^{i \xi t} d\xi. \tag{10.3}$$

Equation (10.3) is the *Fourier inversion formula*: it shows how to recover the function from its Fourier transform. If we define the *inverse Fourier transform* as

$$\check{g}(t) = \frac{1}{2\pi} \int_{-\infty}^{\infty} g(\xi) e^{i \xi t} d\xi, \tag{10.4}$$

we can state Fourier inversion elegantly, as $f = (\widehat{f})^{\vee}$.

We motivated (10.3) by considering a continuous function supported inside $[0, a]$. As shown in the references cited in the Comments and Bibliography section, the Fourier inversion formula (10.3) holds for a much more general class of functions. In particular, it holds for piecewise continuous functions vanishing outside an interval and for functions that rapidly decrease at infinity.

Let us test the Fourier inversion formula for

$$f_1(t) = \begin{cases} 1, & 0 < t < 1, \\ 1/2, & t = 0, 1, \\ 0, & \text{otherwise}. \end{cases} \tag{10.5}$$

The Fourier transform of (10.5) is

$$\widehat{f_1}(\xi) = \int_0^1 e^{-i \xi t} dt = \frac{1 - e^{-i \xi}}{i \xi}. \tag{10.6}$$

According to (10.3),

$$f_1(t) = \int_{-\infty}^{\infty} \frac{1 - e^{-i \xi}}{i \xi} e^{i \xi t} d\xi. \tag{10.7}$$

Symbolic computation of (10.7) is an interesting exercise which we have to forego, as it would take us too deep into complex analysis. Instead, we will compute (10.7) numerically, which is also nontrivial since the integral is improper and the integrand is highly oscillatory for large t.

First we define the function (10.5) and its Fourier transform (10.6) as subfunctions.

```
function y = f(t)
y = double((t>0) & (t<1));
y(t==0) = .5; y(t==1) = .5;
end


fhat = @(xi) (1-exp(-1i*xi))./(1i*xi);
```

The MATLAB quadrature routine, appropriate for integrals such as (10.7), is quadgk— the adaptive 17-point Gauss–Kronrod quadrature. It can be called with infinite limits of integration, however, since the integrand in (10.7) decays relatively slowly, we replaced infinite limits with finite limits.

```
t = linspace(-2,4,400); y = zeros(size(t));

figure;
for n=1:4
    T = 20*n;
    for k=1:length(t)
        g = @(xi) fhat(xi).*exp(1i*xi*t(k));
        y(k) = quadgk(g,-T,T)/(2*pi);
    end
    subplot(2,2,n); hold on;
    plot(t,real(y),'k-'); plot(t,f(t),'k-');
    xlim([-2 4]); ylim([-.5 1.5]);
    text(3,.75,sprintf('T=%g',T))
end
```

The results are shown in Fig. 10.2. It is clear that

$$\lim_{T \to \infty} \int_{-T}^{T} \frac{1-e^{-i\xi}}{i\xi} e^{i\xi t} d\xi = f_1(t)$$

in accordance with (10.3). Notice that at $t = 0, 1$ we get the Fourier value of f_1.

Next, let us compute the Fourier transform and test the Fourier inversion formula (10.3) for the gaussian

$$f_2(t) = e^{-at^2}. \tag{10.8}$$

The integral

$$\widehat{f_2}(\xi) = \int_{-\infty}^{\infty} e^{-at^2} e^{-i\xi t} dt \tag{10.9}$$

can be computed symbolically, using the following trick. Differentiate (10.9) with respect to ξ; on the right-hand side, bring differentiation inside the integral and use integration by parts to relate the resulting integral to $\widehat{f_2}$:

10.1 Fourier Transform

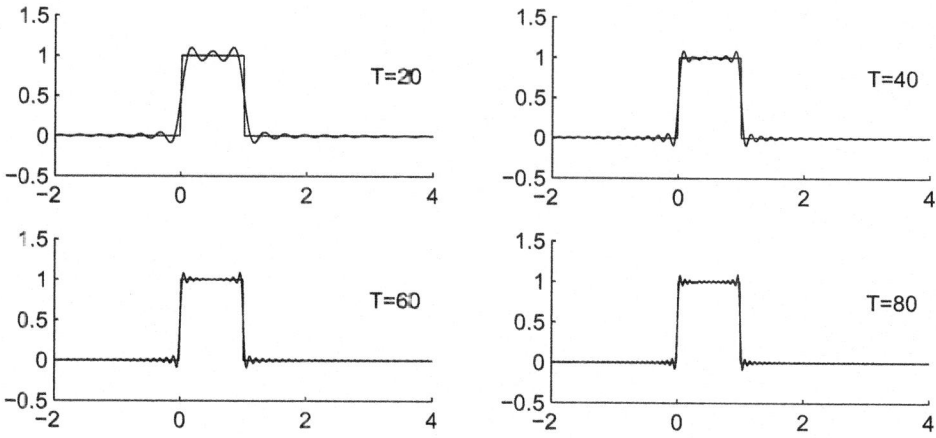

Fig. 10.2 Test of Fourier inversion formula (10.3) for the function (10.5)

$$\frac{d\widehat{f_2}}{d\xi} = \frac{d}{d\xi} \int_{-\infty}^{\infty} e^{-at^2} e^{-i\xi t} dt = \int_{-\infty}^{\infty} e^{-at^2} (-it) e^{-i\xi t} dt$$

$$= \frac{i}{2a} \int_{-\infty}^{\infty} \underbrace{e^{-i\xi t}}_{u} \underbrace{e^{-at^2}(-2at) dt}_{dv}$$

$$= \frac{i}{2a} \left(e^{-i\xi t} e^{-at^2} \Big|_{t=-\infty}^{t=\infty} + i\xi \int_{-\infty}^{\infty} e^{-at^2} e^{-i\xi t} dt \right) = -\frac{\xi}{2a} \widehat{f_2}.$$

Evidently, $\widehat{f_2}$ satisfies a separable ODE

$$\frac{d\widehat{f_2}}{d\xi} = -\frac{\xi}{2a} \widehat{f_2}$$

and is therefore given by $\widehat{f_2}(\xi) = C e^{-\xi^2/4a}$, where

$$C = \widehat{f_2}(0) = \int_{-\infty}^{\infty} e^{-at^2} dt. \tag{10.10}$$

For computing the constant (10.10) there is another trick, which the reader may recognize from multivariate calculus. Write

$$C^2 = \int_{-\infty}^{\infty} e^{-ax^2} dx \int_{-\infty}^{\infty} e^{-ay^2} dy = \int_{-\infty}^{\infty} \int_{-\infty}^{\infty} e^{-a(x^2+y^2)} dx\, dy$$

and switch to polar coordinates:

$$C^2 = \int_0^{2\pi} \int_0^{\infty} e^{-ar^2} r\, dr\, d\theta = \frac{\pi}{a}.$$

Hence $C = \sqrt{\pi/a}$ and the Fourier transform of the gaussian (10.9) is another gaussian

$$\widehat{f_2}(\xi) = \sqrt{\frac{\pi}{a}}\, e^{-\frac{\xi^2}{4a}}. \tag{10.11}$$

A moment of reflection will convince the reader that Fourier inversion takes (10.11) back to (10.9)

$$\frac{1}{2\pi} \int_{-\infty}^{\infty} \sqrt{\frac{\pi}{a}}\, e^{-\frac{\xi^2}{4a}} e^{i\xi t}\, d\xi = e^{-at^2}.$$

This is further confirmed by the following code which produces Fig. 10.3. Notice that we used quadgk with infinite limits: this works because Gaussians are (very) rapidly decreasing.

```
a = 1; f = @(t) exp(-a*t.^2);
fhat = @(xi) sqrt(pi/a)*exp(-xi.^2/4/a);

t = linspace(-3,3);
y = zeros(size(t));
for k = 1:length(t)
    g = @(xi) fhat(xi).*exp(-1i*xi*t(k));
    y(k) = quadgk(g,-inf,inf)/(2*pi);
end

figure
plot(t,f(t),'k-',t,real(y),'k.')
xlabel('t'); ylim([0 1.2]); legend('exact','Fourier');
```

We will now summarize three properties of the Fourier transform that will be useful later.

Let $f_\tau(t) = f(t - \tau)$. In Fourier domain, translation corresponds to multiplication by a complex exponential:

$$\widehat{f_\tau} = \int_{-\infty}^{\infty} f(t-\tau)\, e^{-i\xi t}\, dt = e^{-i\xi\tau} \int_{-\infty}^{\infty} f(s)\, e^{-i\xi s}\, ds = e^{-i\xi\tau} \widehat{f(t)}. \tag{10.12}$$

This is the *shifting property*.

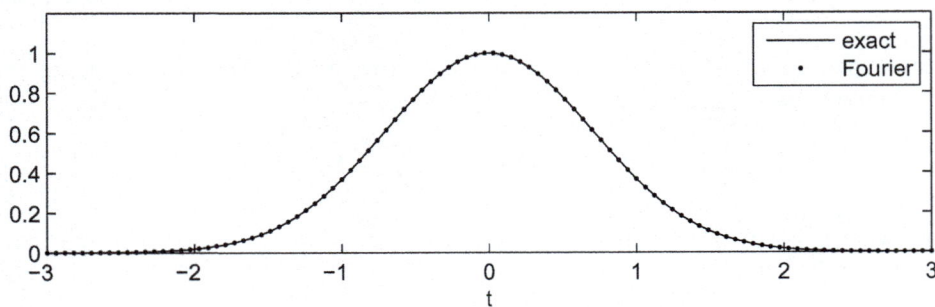

Fig. 10.3 Test of Fourier inversion formula (10.3) for the gaussian (10.9)

The next property is key to Fourier analysis of ODE. Under Fourier transform, differentiation corresponds to multiplication by $(i\xi)$:

$$\widehat{\frac{df}{dt}} = \int_{-\infty}^{\infty} e^{-i\xi t}\, df = e^{-i\xi t} f \Big|_{t=-\infty}^{t=\infty} + (i\xi) \int_{-\infty}^{\infty} e^{-i\xi t} f(t)\, dt \qquad (10.13)$$
$$= (i\xi)\, \widehat{f}.$$

An immediate consequence of (10.13) is that $f^{(n)}$ is transformed into $(i\xi)^n\, \widehat{f}$.

Finally, Fourier transform has a *convolution property*. The convolution of functions on the real line is defined differently from the convolution on the half-line which we discussed in Sect. 7.5, however the notation for the operation is the same:

$$(f * g)(t) = \int_{-\infty}^{\infty} f(s)\, g(t-s)\, ds. \qquad (10.14)$$

Notice that the difference between (10.14) and (7.39) is in the limits of integration. Fourier transform maps convolution into a product:

$$\widehat{f * g} = \int_{-\infty}^{\infty} e^{-i\xi t} \left[\int_{-\infty}^{\infty} f(s)\, g(t-s)\, ds \right] dt$$
$$= \int_{-\infty}^{\infty} f(s) \left[\int_{-\infty}^{\infty} e^{-i\xi t} g_s(t)\, dt \right] ds \qquad (10.15)$$
$$= \int_{-\infty}^{\infty} f(s)\, e^{-i\xi s}\, \widehat{g}(\xi)\, ds = \widehat{f}\, \widehat{g}.$$

Notice the use of the shifting property (10.12).

10.2 Frequency Response Analysis

So far, we have been working mostly with initial value problems. When ODE are analyzed using Fourier transform, time "starts" at $-\infty$ and there are no initial conditions or, rather, all functions involved in the ODE are required to be rapidly decreasing at both infinities. This actually makes physical sense. For instance, consider an RC-circuit: if the capacitor has some initial charge $Q_0 = Q(0)$, it is because it had been charged prior to $t = 0$ through application of external voltage. Same for a mass-spring system: if a mass has initial displacement, it is because it had been displaced from equilibrium with external force. In the following ODE, all functions are rapidly decreasing functions on the real line and there are no initial conditions, as they are subsumed by forcing terms.

Consider a generic scalar linear nonhomogeneous ODE with constant coefficients:

$$a_n x^{(n)} + a_{n-1} x^{(n-1)} + a_{n-2} x^{(n-2)} + \cdots + a_0 x = f. \tag{10.16}$$

By linearity of the Fourier transform and transformation property (10.13), the Fourier transform of the left-hand side is the product $p(i\,\xi)\,\widehat{x}$ where p is the characteristic polynomial of (10.16). Accordingly, as follows from the Fourier inversion formula (10.3)

$$x = \frac{1}{2\pi} \int_{-\infty}^{\infty} \frac{\widehat{f}(\xi)}{p(i\,\xi)} e^{i\,\xi\,t} d\xi. \tag{10.17}$$

We could also derive (10.17) by taking the limit of the Fourier series solution (9.25).

Solving matrix-vector systems with Fourier transform is very similar. Transforming the first order system

$$\frac{d\mathbf{x}}{dt} = A\,\mathbf{x} + \mathbf{f} \tag{10.18}$$

leads to the relationship $(i\,\xi)\,\widehat{\mathbf{x}} = A\,\widehat{\mathbf{x}} + \widehat{\mathbf{f}}$ whence follows that

$$\mathbf{x} = \frac{1}{2\pi} \int_{-\infty}^{\infty} (i\,\xi\,I - A)^{-1}\widehat{\mathbf{f}}\,e^{i\,\xi\,t} d\xi. \tag{10.19}$$

The Fourier solutions (10.17) and (10.19) can be expressed as convolutions. As follows from (10.15), Eq. (10.17) may be written as $x = G * f$ where convolution is defined by (10.14) and

$$G(t) = \frac{1}{2\pi} \int_{-\infty}^{\infty} \frac{1}{p(i\,\xi)} e^{i\,\xi\,t} d\xi. \tag{10.20}$$

Likewise, (10.19) can be expressed as multidimensional convolution $\mathbf{x} = G * \mathbf{f}$ with

$$G(t) = \frac{1}{2\pi} \int_{-\infty}^{\infty} (i\,\xi\,I - A)^{-1} e^{i\,\xi\,t} d\xi. \tag{10.21}$$

As a concrete illustration, let us revisit the 1DOF mass-spring system from Sect. 7.2

$$m\ddot{x} + r\dot{x} + kx = f. \tag{10.22}$$

Instead of specifying zero initial conditions, we stipulate that the forcing term f in (10.22) is zero for negative times: this ensures that at time zero the system is quiescent. For ODE (10.22) Eq. (10.20) becomes

$$G(t) = \frac{1}{2\pi} \int_{-\infty}^{\infty} \frac{1}{m\,\xi^2 + r\,i\,\xi + k} e^{i\,\xi\,t} d\xi. \tag{10.23}$$

10.2 Frequency Response Analysis

The following code shows that (10.23) is consistent with Eq. (7.43)

```
m = 2; r = .5; k = 2 ; p = [m r k];
lambda = roots(p);
t = linspace(-5,40,200);
G1 = (exp(lambda(2)*t) - exp(lambda(1)*t))/...
    (lambda(2)-lambda(1))/m;
G1(t<0) = 0;
G2 = zeros(size(G1));
for k=1:length(t)
    g = @(xi) exp(1i*t(k)*xi)./polyval(p,1i*xi);
    G2(k) = quadgk(g,-20,20)/2/pi;
end
figure; plot(t,G1,'k-', t,real(G2),'ko')
xlabel('t'); ylabel('G'); legend('exact','Fourier')
```

According to Fig. 10.4, for $t \geq 0$ the Fourier integral (10.23) is

$$G(t) = \frac{1}{m} \frac{e^{\lambda_2 t} - e^{\lambda_1 t}}{\lambda_2 - \lambda_1}$$

and is zero otherwise. Since both $G(t)$ and $f(t)$ vanish for $t < 0$

$$x(t) = \int_{-\infty}^{\infty} G(t-s) f(s)\,ds = \int_0^t G(t-s)\, f(s)\,ds$$

which matches (7.43). Equations (10.20) and (10.21) are therefore expressions for impulse response.

Whereas in time domain the relationship between the solution of an ODE and the forcing term is complicated, in Fourier domain it is given by simple multiplication. In the case of the scalar ODE (10.16), the Fourier transform of the solution is obtained by multiplying the Fourier transform of the forcing term by the reciprocal of the characteristic polynomial evaluated at $i\xi$. Thus all of the information about the system modeled by (10.16) is encoded in the function $h(\lambda) = 1/p(\lambda)$. Similarly, the matrix-valued function $H(\lambda) = (\lambda I - A)^{-1}$

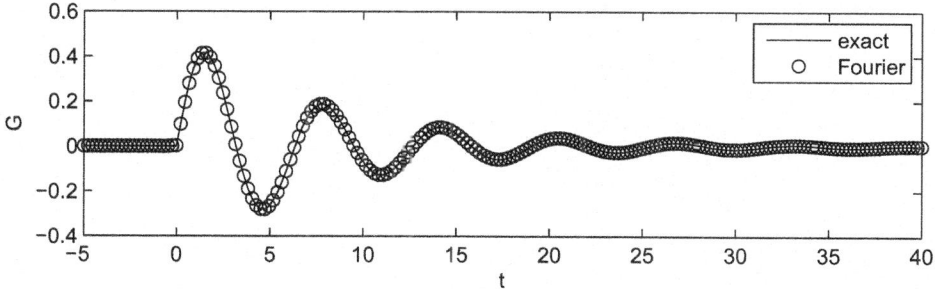

Fig. 10.4 Computation of impulse response for 1DOF mass-spring system using (10.23)

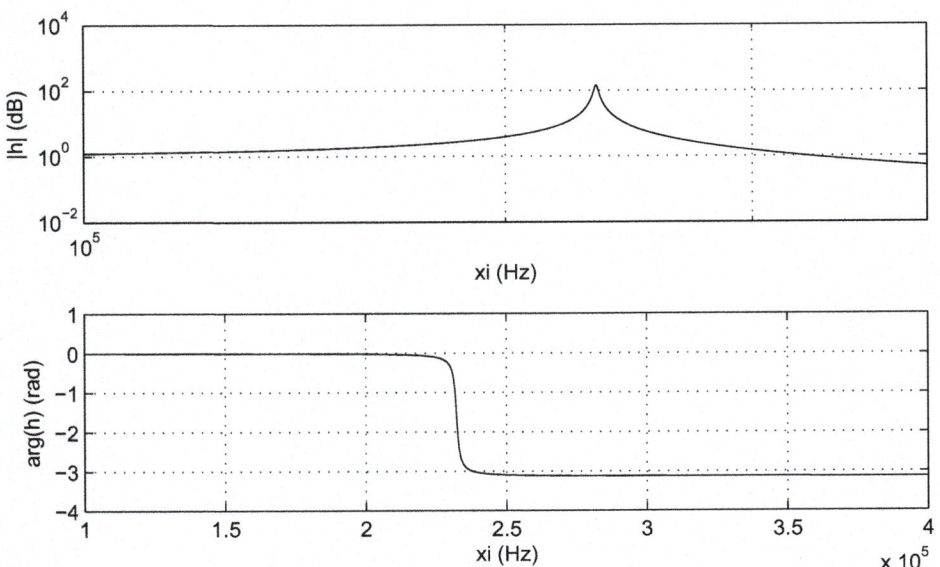

Fig. 10.5 Bode diagram for the RLC-circuit in Fig. 7.2 with $R = 10\,\Omega$, $L = 1\,\text{mH}$, and $C = 470\,\text{pF}$

encodes all of the necessary information for the system modeled by (10.18). These are examples of *transfer functions*.

When a physical system is modeled with a linear ODE with constant coefficients, the goal is usually to determine its transfer function. The value of the transfer function at $i\,\xi$ determines the response of the system to the complex exponential of that frequency, and conversely. For that reason, determination of transfer functions is called *frequency response analysis*.

One way to visualize a transfer function $h(\lambda)$ is by plotting $h(i\,\xi)$ in the complex plane. This was the method in Sect. 9.4 where we investigated an RC-circuit driven by a square waveform (Figs. 9.4 and 9.7). A more common way to visualize a transfer function is with a *Bode diagram* which consists of the log-log plot of $|h(i\,\xi)|$ and a regular plot of $\arg(h(i\,\xi))$ against ξ.

The following code creates Bode diagram for the RLC-circuit in Fig. 7.2 with nominal parameters $R = 10\,\Omega$, $L = 1\,\text{mH}$, and $C = 470\,\text{pF}$ (Fig. 10.5).

```
R = 10; L = 1e-3; C = 470e-12;
xi = 1e5:100:4e5;
h = 1./polyval([L*C R*C 1], 2*pi*1i*xi);

figure; subplot(2,1,1);
loglog(xi,abs(h),'k-'); grid on;
xlabel('xi (Hz)'); ylabel('|h| (dB)');
```

```
subplot(2,1,2);
plot(xi,angle(h),'k-'); grid on;
ylim([[-4 1]])
xlabel('xi (Hz)'); ylabel('arg(h) (rad)');
```

The plot in the top panel has a peak near 232 kHZ. This is the *resonant frequency*: the response of the circuit (voltage across the capacitor) is greatest if the frequency of the driving voltage equals the resonant frequency. The bottom plot shows how harmonic components of the driving voltage are shifted as they "pass" through the RLC-circuit. Low frequency components are practically unshifted, but high frequency components—with frequencies greater than the resonant frequency—are shifted by π.

As a more sophisticated example, consider the train of two carts in Fig. 10.6; this is a 2DOF mass-spring system. Suppose that external force f is applied to m_2 and the displacement x_1 is measured as response. Let us find the transfer function from f to x_1 which is the ratio $\widehat{x}_1/\widehat{f}$.

As follows from Newton's second law, the scalar equations of motion are

$$m_1 \ddot{x}_1 = -k_1 x_1 + k_2 (x_2 - x_1) - r \dot{x}_1,$$
$$m_2 \ddot{x}_2 = -k_2 (x_2 - x_1) - k_3 x_2 - r \dot{x}_2 = f. \tag{10.24}$$

In order to rewrite (10.24) in matrix-vector form, introduce vectors $\mathbf{x} = \begin{bmatrix} x_1 & x_2 \end{bmatrix}^T$, $\mathbf{b} = \begin{bmatrix} 0 & 1 \end{bmatrix}^T$, and matrices

$$M = \begin{bmatrix} m_1 & 0 \\ 0 & m_2 \end{bmatrix}, \quad R = \begin{bmatrix} r & 0 \\ 0 & r \end{bmatrix}, \quad K = \begin{bmatrix} k_1 + k_2 & -k_2 \\ -k_2 & k_2 + k_3 \end{bmatrix}.$$

Now (10.24) can be written as

$$M \ddot{\mathbf{x}} + R \dot{\mathbf{x}} + K \mathbf{x} = f \mathbf{b}. \tag{10.25}$$

Applying the Fourier transform to (10.25) gives the relation

$$\left((i\xi)^2 M - (i\xi) R + K \right) \widehat{\mathbf{x}} = \widehat{f} \mathbf{b}$$

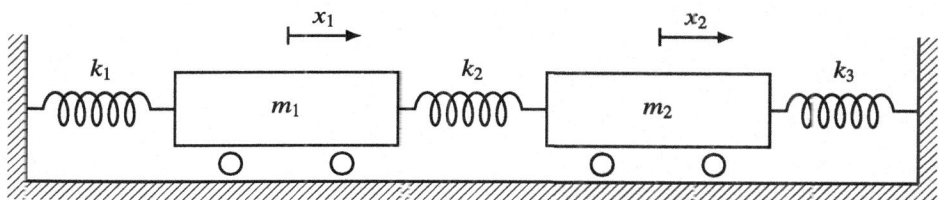

Fig. 10.6 A train of two carts

from which follows that

$$\frac{\widehat{x_1}}{\widehat{f}} = \mathbf{c}^T \left((i\,\xi)^2 M + (i\,\xi) R + K \right)^{-1} \mathbf{b}, \tag{10.26}$$

where $\mathbf{c} = \begin{bmatrix} 1 & 0 \end{bmatrix}^T$. It is instructive to compare the exact expression (10.26) with approximation computed from data. We simulated the latter in MATLAB by solving (10.25) with the following values of the parameters: $m_1 = 2$, $m_2 = 1$, $k_1 = k_3 = 200$, $k_2 = 100$, and $r = 0.5$. The expression for the forcing term

$$f = \sin(0.3\,t^2) \tag{10.27}$$

and the time interval $[0, 50]$ were chosen so that DFT of the forcing term (sampled on the grid) resembles a characteristic function of an interval containing natural frequencies of the mass-spring system.

```
m = [2 1]; r = .5; k = [200 100 200];
M = diag(m); I = eye(size(M)); R = r*I;
K = [k(1)+k(2) -k(2); -k(2) k(2)+k(3)];
A = [zeros(size(M)) I; -M\K -M\R];
b = [0;0;M\[0;1]]; f = @(t) sin(.5*t.^2);
odefun = @(t,u) A*u + f(t)*b;
N = 1024; T = 50; t = T*(0:N-1)/N;
[t,u] = ode45(odefun,t,zeros(4,1));
ft = f(t); x1 = u(:,1);

figure;
subplot(2,1,1); plot(t,ft,'k');
xlabel('t (s)'); ylabel('f(t) (N)');
subplot(2,1,2); plot(t,x1,'k');
xlabel('t (s)'); ylabel('x1(t) (N)');
```

Figure 10.7 shows how, as the forcing term (top panel) becomes more and more oscillatory, the displacement x_1 first gains amplitude, then loses it, then gains it again before losing it for good. This is indicative of two resonant frequencies which are seen as peaks in the top panel of Fig. 10.8.

```
xi = (0:N-1)/T; h1 = fft(x1)./fft(ft);
h2 = zeros(size(xi));
for n=1:length(xi)
    w = 2*pi*1i*xi(n);
    h2(n) = [1 0]*( (w^2*M+w*R+K)\[0;1]);
end

figure;
subplot(2,1,1);
loglog(xi,abs(h1),'k.-',xi,abs(h2),'k-');
xlim([0 10]); grid on;
```

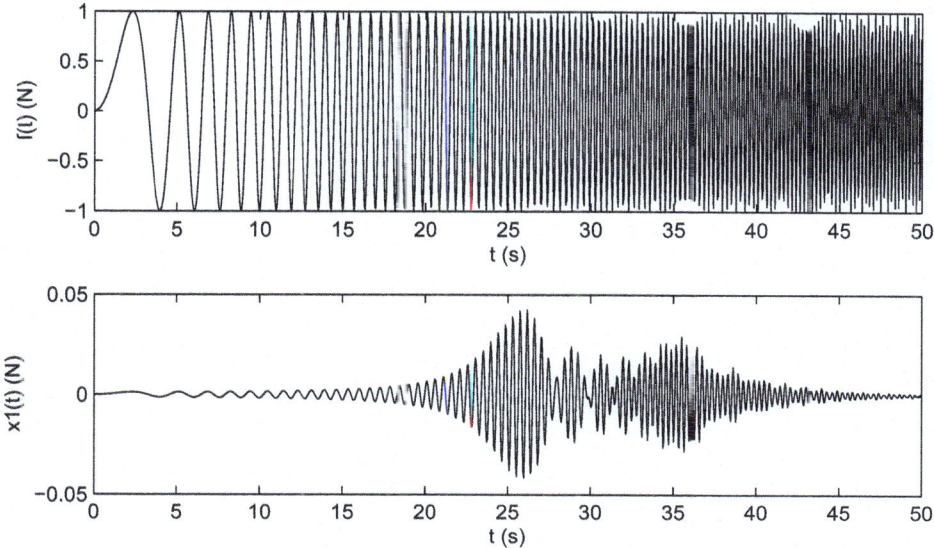

Fig. 10.7 The solution of (10.25) (bottom panel) with forcing term (10.27) (top panel)

```
xlabel('xi (Hz)'); ylabel('|h| (dB)');
subplot(2,1,2);
plot(xi,angle(h1),'k.-',xi,angle(h2),'k-');
xlim([0 10]); grid on;
xlabel('xi (Hz)'); ylabel('arg(h) (rad)');
```

Figure 10.8 shows very good agreement between (10.26) and its DFT estimate for $|\xi| < 4.5$. For $\xi > 4.5$ the Fourier transform of the forcing term becomes small and the errors in the DFT coefficients swamp the computation. This can be corrected by changing the forcing term, however it is not straightforward.

10.3 Heat Equation on a Line

In Chap. 9 we solved the heat equation (9.42) on an interval using Fourier series. Here we will solve it on the real line. Application of the Fourier transform to

$$\frac{\partial u}{\partial t} = \alpha \frac{\partial^2 u}{\partial x^2}, \quad \left(\alpha = \frac{k}{c\rho}\right), \quad -\infty < x < \infty, \tag{10.28}$$

$$u(x, 0) = f(x)$$

gives the IVP

$$\frac{d\widehat{u}}{dt} = -\alpha \xi^2 \widehat{u}, \quad \widehat{u}(0) = \widehat{f}$$

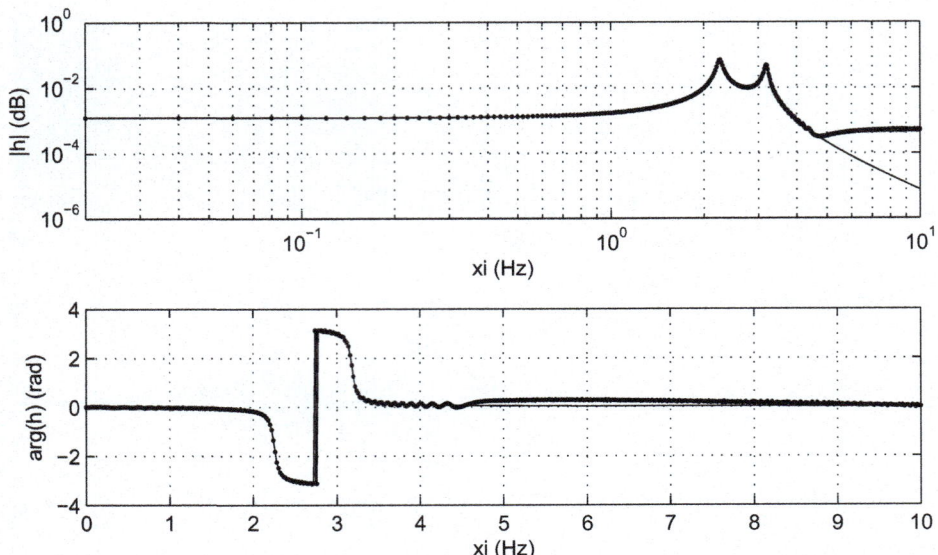

Fig. 10.8 Bode diagram computed from the data in Fig. 10.7 (dotted line) and (10.26) (solid line)

whose solution is

$$\widehat{u} = \widehat{f} e^{-\alpha \xi^2 t}. \tag{10.29}$$

Applying the inverse Fourier transform (10.4) gives the solution of (10.28) in the form:

$$u = \left(\widehat{f} e^{-\alpha \xi^2 t}\right)^{\vee}. \tag{10.30}$$

Since (10.30) is the inverse Fourier transform of a product of Fourier transforms, the result is a convolution. Furthermore, since Fourier transform maps Gaussians into Gaussians (Eq. (10.11)), the solution of (10.28) is convolution of the initial condition with the Gaussian:

$$u(x,t) = \int_{-\infty}^{\infty} \frac{1}{\sqrt{4\pi t}} e^{-\frac{(x-s)^2}{4\alpha t}} f(s)\, ds. \tag{10.31}$$

The expression multiplying f inside the integral in (10.30) is the *heat kernel*—an important object in mathematical analysis.

10.4 Laplace Transform

Laplace transform is defined for functions on the half-line as follows:

$$L(f) = \int_0^{\infty} e^{-st} f(t)\, dt. \tag{10.32}$$

1C.4 Laplace Transform

If the function f is extended by zero for negative times and the parameter s is replaced with $i\xi$, Eq. (10.32) becomes the Fourier transform (10.2). It is therefore unsurprising that Laplace transform has properties that are very similar to those of the Fourier transform. The two main properties, whose proof is left as an exercise, are the transformation of the derivative

$$L\left(\frac{df}{dt}\right) = s L(f) - f(0) \tag{10.33}$$

and the convolution property, where convolution is the same as in Chap. 7:

$$L(f * g) = L(f) L(g), \quad (f * g)(t) = \int_0^t f(t-s) g(s)\, ds. \tag{10.34}$$

The use of Laplace transform is very similar to that of Fourier transform. For instance, consider the 1DOF mass-spring system with zero initial conditions:

$$m\ddot{x} + r\dot{x} + kx = f, \quad x(0) = \dot{x}(0) = 0.$$

Application of (10.32) leads to

$$(m s^2 + r s + k) L(x) = L(f),$$

whence follows that

$$L(x) = \frac{1}{p(s)} L(f), \quad p(s) = m s^2 + r s + k. \tag{10.35}$$

It so happens that Laplace transforms of elementary functions are rational functions:

$$L(t^n) = \frac{1}{s^{n+1}}, \quad L(e^{at}) = \frac{1}{s+a}, \quad L(\cos(\omega t)) = \frac{s}{s^2 + \omega^2}, \quad \text{etc.}$$

If the forcing term f is an elementary function, the inverse Laplace transform of (10.35) can be decomposed into partial fractions and identified using a table of Laplace transforms. However, if f is not elementary then the inversion of the Laplace transform must be performed via Fourier inversion.

Equation (10.35) immediately gives the transfer function

$$h(s) = \frac{L(x)}{L(f)} = \frac{1}{p(s)}.$$

It may be a bit more expedient to compute transfer functions as ratios of Laplace transforms rather than ratios of Fourier transforms. However, as convenient as Laplace transform may be in certain cases, Fourier transform is foundationally more important.

10.5 Comments and Bibliography

Fourier transform and its applications to ODE is the main topic of [2]. Fourier analysis of linear PDE with constant coefficients is discussed in most introductory PDE books, such as [3].

Figure 10.8 shows that computing transfer functions is nontrivial even for simulated data; for real data it can be very complicated. We did not include our experimental measurement of the transfer function for the RLC-circuit used as the first example in Sect. 10.2 because we got a noisy peak that was not only much lower than in Fig. 10.4 but also shifted by 20 kHZ. The large discrepancy is due only in part to the tolerances for the values of R, L, and C—the main culprit is capacitive loading from the probes of the oscilloscope. Mass-spring systems have lower natural frequencies and are not susceptible to capacitive loading. However, our physical train of two carts turned out to have nonlinear friction. Also, we could not control the forcing term precisely enough.

Laplace transform was popularized in electrical engineering by Oliver Heaviside in early 1880s and is still a weapon of choice among engineers. While it is good for quickly deriving symbolic solutions and transfer functions of simple ODE, it is not nearly as useful as the Fourier transform. In fact, computationally, inverting Laplace transform is a nightmare—see [1].

10.6 Exercises

1 Use Fourier transform to solve the following ODE with a delay term:

$$\frac{dx}{dt} + x + x(t-1) = f.$$

Validate the solution numerically for several choices of f.

2 The transfer function from f to V for the RC-circuit modeled by

$$RC\frac{dV}{dt} + V = f$$

is given by

$$h(\lambda) = \frac{1}{1 + RC\lambda}.$$

Confirm this by computing the transfer function from simulated (or real) data using DFT, as we did in Sect. 10.2.

3 Find the transfer function for a train of N carts and show that, for small friction, its Bode plot has N resonant peaks.

4. Solve the wave equation on the line:

$$\frac{\partial^2 u}{\partial t^2} = \frac{1}{c^2}\frac{\partial^2 u}{\partial x^2}, \quad -\infty < x < \infty, \quad t > 0,$$

$$u(x,0) = f(x), \quad \frac{\partial u}{\partial t}(x,0) = g(x). \tag{10.36}$$

Plot the solution for several values of time (or animate it) with $g = 0$ and

$$f(x) = \begin{cases} 1 - |x|, & -1 \le x \le 1, \\ 0, & \text{otherwise.} \end{cases}$$

What is the physical interpretation of the plots?

5. Consider the heat equation (10.28). It should be solved using Fourier transform, as demonstrated in Sect. 10.3, but it can also be solved using Laplace transform, at least in theory. Let $\widetilde{u}(x,s) = \int_0^\infty e^{-st} u(x,t)\, dt$ be the Laplace transform of $u(x,t)$. Show that it satisfies

$$\alpha \frac{d^2 \widetilde{u}}{dx^2} - s\widetilde{u} = -f.$$

Solve this ODE and try to show that its solution is the Laplace transform of (10.31); this is a difficult exercise.

References

1. C.L. Epstein, J.C. Schotland, The bad truth about laplace's transform. SIAM Rev. **50**(3), 504–520 (2008)
2. C. Gasquet, R. Ryan, P. Witomski, *Fourier Analysis and Applications: Filtering, Numerical Computation, Wavelets*. Texts in Applied Mathematics (Springer, New York, 1998)
3. W.A. Strauss. *Partial Differential Equations: An Introduction* (Wiley, 2007)

Index

A
Abelian group, 72
Affine space, 92

B
Barometric law, 39
Basis
    and linear independence, 74
    components with respect to, 74
    orthogonal, 126
    representation, 74
Biot number, 29
Biot's law, 56
Boundary condition
    Dirichlet, 161, 167
    Neumann, 68, 161, 163, 164
    periodic, 161

C
Carrying capacity, 7
Characteristic altitude, 39
Characteristic function, 112
Characteristic polynomial, 100, 103, 105–107, 121, 151, 176, 177
Complimentary function, 81
Conformal plot, 63
Contravariant tensors, 90
Convection heat transfer coefficient, 28
Convolution, 98, 109, 114, 115, 124, 169, 175, 176, 182, 183

D
Dirac delta function, 122
Division algebras, 90

E
Eigendecomposition, 58, 66, 97–101, 103, 104, 117, 121
Eigenfunction, 121
Eigenvalue, 99
Eigenvector, 99
Equation of continuity, 161
Euclidean space, 72
Euler's formula, 63

F
Fick's law, 43
Fourier inversion formula, 171
Fourier series, 143–145, 147, 148, 150–154, 159, 161–165
    lacunary, 165
Fourier's law, 161, 166
Fourier transform, 169–172, 174–177, 179, 181–184
Fourier value, 143, 145, 148, 165
Fourier's law, 41, 42

G
Gibbs phenomenon, 129, 147, 148

H
Heat equation, 68, 160, 161, 163, 164, 166, 169, 181
Hilbert spaces, 165

I
Identity matrix, 80
Impulse response, 98, 117–119, 121–123, 177
Initial value problem, 5
Inner product, 72
    complex, 130, 131, 144, 150
    dot product, 148
    geometry, 149
    standard L^2, 167
    weighted L^2, 150

K
Kirchhoff's current law, 33
Kirchhoff's voltage law, 33

L
Lambert law, 53
Laplace transform, 64, 182
Law of mass action, 10
Likelihood, 14
Linear operator, 77
Logistic equation, 6

M
Machine epsilon, 14
Malthusian catastrophe, 10
Mass-spring system
    1DOF, 102
    2DOF, 179
    driven by harmonic force, 106
    electro-mechanical analogy, 105
    equivalent circuit, 104
    natural frequency, 108
Matrix exponential, 65
Matrix representation, 76
Method of undetermined coefficients, 88, 92, 98, 106, 108, 109, 115, 118, 124, 151, 152
Mutarotation of glucose, 56

N
NASA 1976 standard atmosphere model, 39
Natural growth equation, 2
Nelder-Mead simplex method, 9
Newton's law of cooling, 29
Newton's method
    basins of attraction, 18
    cycle, 17
    multidimensional, 20
    one-dimensional, 16
    quadratic convergence, 18
Norm, 5, 11, 72
Null space, 77
Number field, 70

O
Octonions, 90
Ohm's law, 33

P
Principle of superposition, 69, 86, 87, 107, 112, 138, 151–153, 164
Probe loading, 35

Q
Quoternions, 89

R
Radioactive decay
    as a memoryless process, 49
    Markov chain, 52
Range, 77
RC-circuit
    driven by a simple harmonic, 83
    driven by square waveform, 153
    electric, 32–34, 43
    electro-thermal analogy, 40
    hydraulic, 31
    loaded with a probe, 46
Resonance, 108
Riccati equation, 92
Ring, 70
RLC-circuit
    Bode diagram, 178

Index

step and impulse response, 118

S
Separation of variables, 3
Soakage, 35
Stefan-Boltzmann law, 42
Step response, 98, 118–120
Structure theorem, 69, 81, 83, 86, 97

T
Terminal velocity, 36
Thermal diffusivity, 161

Time constant, 28
Transient, 86

U
Usain Bolt's world record, 37

V
Variation of parameters, 11
Vector space, 71
Volta's law, 33
Volterra integral equation, 80

SPRINGER NATURE

GPSR Compliance

The European Union's (EU) General Product Safety Regulation (GPSR) is a set of rules that requires consumer products to be safe and our obligations to ensure this.

If you have any concerns about our products, you can contact us on ProductSafety@springernature.com

In case Publisher is established outside the EU, the EU authorized representative is:

Springer Nature Customer Service Center GmbH
Europaplatz 3
69115 Heidelberg, Germany

The manufacturer's authorised representative in the EU is Springer Nature Customer Service Centre GmbH, Europaplatz 3, 69115 Heidelberg, Germany. If you have any concerns regarding our products, please contact ProductSafety@springernature.com

Printed and bound by CPI Group (UK) Ltd, Croydon, CR0 4YY
05/01/2026
02029247-0001